POSTTRAUMATIC STRESS DISORDER

The Victim's Guide To Healing and Recovery

Raymond B. Flannery, Jr., Ph.D.

CHEVRON

PUBLISHING CORPORATION

2004
Chevron Publishing Corporation
5018 Dorsey Hall Drive, Suite 104
Ellicott City, MD 21042
(410) 740-0065

1992
The Crossroad Publishing Company
370 Lexington Avenue
New York, NY 10017

ISBN: 978-1-935307-46-4

For Georgina G. Maybury
and
Victims Everywhere

Contents

Preface to the Chevron Edition

A decade ago, this victim's guide was the first book to translate the technical medical and scientific findings of posttraumatic stress disorder (PTSD) into simple, understandable language so that victims could understand what was happening to them and how to overcome it. It was eagerly received by persons who knew that they were victims and by counselors who wanted to help them. Often, both patient and counselor would each have a copy of the book to use as a common frame of reference in their work together.

As word of the book spread, individuals who knew something in their lives was wrong but couldn't identify it read the book and many realized that they too were victims of untreated PTSD. Equally encouraging, they had a sense of hope that they could do something to reduce their distress.

The book gathered an additional set of readers during the events of Sept. 11, 2001, at the World Trade Center in New York City, as lay people everywhere understood from personal experience the impact of psychological trauma and PTSD.

During the past decade, the world has experienced several traumatic events. Wars in Kuwait, the former Yugoslavia, Afghanistan, and Iraq have resulted in combat veterans with PTSD. The rise of terrorism and events like the World Trade Center and Bali have produced similar victims with PTSD. Natural and man-made disasters, including the space shuttle *Columbia*, have occurred, as has man's enduring legacy of violence in the forms of murder, rape, robbery, and aggravated assault. Sadly, the world of victimization continues.

However, also during this past decade, medical and behavioral scientists have continued to work quietly to further our understanding of the biology and psychology of psychological trauma and posttraumatic stress disorder. As was true ten years ago, much of this information is still only found in highly technical professional journals.

This second edition again incorporates these latest findings into our body of knowledge of psychological trauma and posttraumatic stress disorder into workable language for victims. Included are refinements in diagnosis, biochemistry, risk factors, and untreated consequences. These serve to strengthen the book's basic understanding of PTSD and the steps needed to resolve its aftermath.

My hope for victims in this revised edition is the same as it was ten years ago: that this book may help to relieve your journey into the heart of darkness and that your long night of suffering may finally come to an end.

—Raymond B. Flannery, Jr., PhD, FAPM
Autumn, 2003

Preface

There have been many important recent advances in our understanding and treatment of the severe life stress known as *Posttraumatic Stress Disorder* (PTSD). These findings have been largely unavailable to the general public, and this book addresses that need. It is especially written for victims who are suffering from the untreated consequences of posttraumatic stress disorder and for persons with addictive behavior, many of whom are self-medicating the untreated effects of PTSD without realizing it.

Posttraumatic stress disorder (PTSD) grows out of events that can cause severe life stress in any one of us. These events, known as psychological trauma, are overwhelming in their impact on us, and we have no control over them no matter how hard we try. Sexual abuse, physical abuse, combat, and major disasters are common examples of these events. Traumatic events may disrupt our sense of reasonable mastery, alter our attachments to others, and leave us unable to make any sense of what is happening to us. If left untreated, these disruptions along with physical symptoms such as being unable to relax, being easily startled, or having recurring nightmares or intrusive memories form the medical condition known as posttraumatic stress disorder (PTSD).

Addictive Behaviors are compulsive patterns of behavior or of drug usage, and may include the use of alcohol and drugs like crack/cocaine, sexual addictions, gambling, some cases of sensation seeking, high risk-taking behavior, repetitive self-mutilation, and so forth. Recent medical evidence has shown that many of these persons with addictive behaviors are also victims of posttraumatic stress disorder, and that the addictive behavior appears to be an attempt to self-medicate the physical and mental distress that arises when the painful effects of psychological trauma and PTSD are left untreated.

This book will help you understand what posttraumatic stress disorder is, what its signs of distress are, and how it is linked to addictive behavior in many victims. It will help you determine if you have

PTSD. If you do, this book will then help you learn what feelings your traumatic event(s) has (have) stirred in you, and what you can do to recover successfully and get on with your life.

How To Use This Book. The first section of this book presents a state-of-the-art review of the general nature of posttraumatic stress disorder. Victims of any type of traumatic event will want to read this section, as this medical condition is highly similar for all victims. Victims with addictive behavior will want to pay particular attention to Chapter 3, which explores the biological basis of addictive behavior in some victims.

The second section discusses our most recent understanding of specific types of traumatic events. Victims of sexual abuse, physical abuse, combat, and family alcoholism will want to read that chapter relevant to their own circumstances. This section may be profitably read, however, by all victims, since the experience of victimization has many communalities, even though specific types of traumatic events may differ.

The final section outlines specific and up-to-date pathways for recovery and healing, and again should be read by each victim. One important aspect of this full recovery is the victim's ability to make meaningful sense of the painful event that has occurred. While medicine and psychology can help us understand some aspects of violence, often the victim's questions go beyond the scientific data. For this reason I have felt it important in the last chapter to explore more fully the problem of making meaningful sense of violent acts. Society has evolved several ways to consider this problem (including scientific understanding), and I have noted at the end some of the more common pathways that have proven helpful to others.

Every chapter begins with a case example which is meant to illustrate particular concepts in that chapter, and also to communicate the universal experience of victimhood. These are true, but they are also very painful. Some victims may find these examples comforting as they learn that they are not alone in what has happened to them. Other victims may find these studies revive painful memories. Each victim must decide whether and when to read these vignettes.

Finally, since the book reports on ground-breaking research, the Select Reading list is more detailed than might ordinarily be the case. I have done this so that victims, their loved ones and friends, and their counselors may explore more fully any aspect of PTSD that might be of special relevance in any given set of circumstances. This list may be especially helpful for counselors. The more we understand the most recent advances in the field, the better are our efforts to reduce the human suffering associated with posttraumatic stress disorder in those who seek our help.

Abraham Lincoln once said that defeat strikes the young the hardest because they least expect it. It has been my clinical experience that posttraumatic stress disorder shows no such age discrimination. Life's defeats can be hard for any of us at any age. These pages are written for you, the victim, so that you will know that there is help in these painful matters, and that with reasonable effort and time your interest and joy in life can be restored.

* * * * *

Every author is indebted to a variety of different sources for the development of his or her own intellectual roots. A book such as this is immediately indebted to the community of scholars in medicine and behavioral science who work quietly and in small steps to understand the world in which we live. I am equally grateful to my patients and students of the past 25 years for what they have taught me about impact of severe stress on human health, and to my colleagues in the Harvard Trauma Study Group for their thoughtful discussions of the complexities of psychological trauma.

My appreciation is also expressed to the following men and women who have more directly influenced the development of this book: Richard Audet, PhD; Audrey DeLoffi, LICSW; Caroline Fish-Murray, EdD; George Everly, Jr., PhD; Thomas Garrett, PhD; Henry Grunebaum, MD; Carol Hartman, RN, DNSc; Mary Harvey, PhD; Leston Havens, MD; Cynthia Heller; Judith Herman, MD; David Hotchkiss, MDiv; Edward Khantzian, MD; Terrence Keane, PhD; Thomas King, SJ, PhD; H. John McDargh, PhD; Jeffery Mitchell, PhD; Deborah Moran, MD; Malkah Notman, MD; Pauline Paggliocca, PhD; Walter Penk, PhD; J. Christopher Perry, MD; Jonathan Perry, PhD; Roger Pitman, MD; Richard Poor; Mollie Schoenberg; Alan Siegel, EdD; Judith Tausch, RN, EdD; Bessel van der Kolk, MD; and James Woods, SJ, EdD.

All of these men and women have provided thoughtful and wise suggestions. Any errors, however, remain my sole responsibility.

xi

This book is dedicated to my great aunt Georgina who, upon the death of her sister, became my grandmother, and thus by both word and example taught me to be concerned for the welfare of others. This book is equally dedicated to all the victims of violence everywhere so that in some small way the suffering of their present darkness may be lessened.

—Raymond B. Flannery, Jr., PhD, FAPM
Autumn, 1992

Author's Note and Editorial Method

Psychological trauma is a rapidly expanding area of scientific and medical inquiry. As with any aspect of health care, medicines and other forms of treatment are constantly being improved, and this book is not intended to be a substitute for the medical advice of your physician or the counseling of your therapist. Everyone following the suggestions in this book is advised to begin with a physical exam. The reasons are threefold. You want to be sure there are no untreated medical consequences from your past abuse. You want to know that your psychological trauma symptoms are not due to some other medical illness, and lastly, you want to obtain medical clearance for the aerobic exercises spoken of in these pages. Note carefully the specific suggestions and warnings for the aerobic exercise and relaxation components of the Project SMART program presented in this book. Raise any questions you may have with your physician or your counselor, and always following his or her advice first.

The Select Readings list, which has been provided at the end of the volume for further reading, also contains all of the citations noted in the book.

* * * * *

All of the case examples in this book have been disguised to protect the anonymity of those involved.

Part 1

POSTTRAUMATIC STRESS DISORDER: ITS GENERAL NATURE

1

What Has Befallen Me?
Psychological Trauma and
Posttraumatic Stress Disorder

I a stranger and afraid
In a world I never made.
— A.E. Housman

Help me make it through the night.
— Daniel Berrigan

Mary Ellen picked up the bread and headed for the checkout counter. As she stood in line and flipped through a magazine, she was suddenly seized by a sharp, stabbing pain in her chest. A sense of complete desolation followed. It was one o'clock in the afternoon.

"Joey, Joey. Get your head off of the desk. This is not nap time." Second graders, thought their teacher, Mrs. Franklin, could really try one's patience. Gently she nudged Joey to wake him up. Joey did not move. Mrs. Franklin raced to the principal's office to summon emergency aid. The call for the EMTs was recorded at twelve-forty-five in the afternoon. Fifteen minutes later Joey was pronounced dead.

The neighbors had known Mary Ellen as a shy, attractive single parent who worked as a bank teller, volunteered at a senior citizen's center, and who idolized the ground her son Joey walked on.

Hers had seemed the ordinary life of a single parent until his death this day.

The next afternoon Mary Ellen stood numbly before the casket containing her little boy. Her own father tried to console her, but no tears would come. She thought about Joey's birth just a few short years before. She had returned from college when she was pregnant with Joey. Her family and friends were surprised since she had never really shown interest in boys. Swallowing her pride and her shame, she explained that it has all been a youthful fling. Her father was outraged.

After the child's birth, Mary Ellen left college, obtained a job, and embarked upon a career in single parenting. In some ways Joey did not change her life. The hypervigilance, the recurring insomnia, the self-loathing, the sadness—the side of Mary Ellen the community did not see—continued to be her daily companions in her parenting years just as they had been in her childhood.

"Shall we contact Joey's biological father?" The question jarred her. Her eyes returned to the casket, and she held the now still little hand in hers. The tears would still not come. Her own repeated incest at the hands of her father, her own forcible rape in the college campus library, and now the death of her son. What would become of her? Had God forsaken her?

*   *   *

Yearly. Monthly. Daily. Hourly. Even as you read these words, it is happening. Like Mary Ellen, some one of us, somewhere, is becoming a victim of violence, a victim of psychological trauma, either by direct act or by witnessing such an event. The effects of such a trauma can be immediately overwhelming, and, if left untreated, may evolve into *posttraumatic stress disorder* (PTSD) with its alterations in our sense of control, our relationships to others, our personal physical discomfort, and our sense of purpose in life. Consider the following:

- A seven-year-old girl who is being sexually abused by her father is afraid to tell her mother for fear the family will be destroyed, if she reveals the abuse.
- A young teenage male with two alcoholic parents is unable to go to parties with his friends because the normal excitement of

such events always results in his having insomnia.

- A thirty-seven-year-old woman flees home and comes to the hospital emergency room in complete distress. Her husband, who batters her daily, has threatened to kill her, and is, even now, stalking her on nearby streets.
- A fifty-seven-year-old combat veteran is unable to watch television because each televised episode of any type of violence triggers anew intrusive memories of his battlefield experiences of years ago.
- A seventy-year-old grandmother has been unable to drive a car since a drunken driver killed her granddaughter on the night of the granddaughter's high school prom.

These painful examples represent some of the ways the harmful consequences of posttraumatic stress disorder may continue over time. Sexual abuse, physical abuse, combat and terrorism, violent street crime, homicide, serious car accidents, family alcoholism, natural and man-made disasters, sudden life-threatening illness or death are all forms of psychological trauma that each year claim their share of victims who develop PTSD.

PTSD may result from one episode of violence such as being robbed at gunpoint, from repeated episodes of the same violent act such as in repeated incest by the same family assailant, or from repeated but different types of traumatic events such as in our earlier example of Mary Ellen, who is a victim of incest, of campus date-rape, and the sudden death of her own child.

Such traumatic events are frightening and depressingly common. The exact extent of such violence is difficult to determine because reporting methods differ from one policing or health care agency to another, and because many victims do not report these events at all. A recent and reliable study, however, has estimated that as many as two and one-half to possibly five million Americans may be suffering from the harmful effects of PTSD.

Table 1 reveals the extent of traumatic violence in our own country. While the estimates may not be precise, the list of events does show us the range of traumatic events that may befall any of us in our lives. To this list, we could add one-hundred children die every month from abuse or neglect, that one child is raped every forty-five minutes, that one teenager commits suicide every

TABLE 1

The Estimated Impact of Violence in the United States

5% of the elderly are physically, financially, or verbally abused.

14% of all Americans are victims of crime each year.

14% of all marriages have two spouses who batter each other.

20% of all women have been both battered and sexually abused.

20% of all men have been sexually abused.

25% of all women have been sexually abused.

25% of all women have been battered.

30% of all combat victims have been psychologically traumatized.

30% of all persons with Borderline Personality Disorder have a history of child abuse.

40% of all pregnant women have been battered.

50% of all eating disordered individuals have a past history of abuse.

55% of all family violence occurs in alcoholic homes.

60% of all children hit other children.

80% of all persons with sexual addictions come from alcoholic homes.

80% of all persons with Multiple Personality Disorder have a history of child abuse.

---

fifty-nine minutes, that one person is killed by a drunken driver every twenty-two minutes, that one person is murdered every twenty-six minutes. And these estimated statistics do not include those children and adults that are traumatized by witnessing these events happening to others.

As we can see, it is very likely that any one of us or of our loved ones and friends will be touched by some traumatic event in the course of a lifetime. If the impact of the traumatic event is not dealt with immediately (and much trauma is not), we or they remain at high risk to develop posttraumatic stress disorder.

Philosopher Francis Bacon wrote long ago that knowledge itself is power. The goal in this chapter and in this book is to study what befalls a person who becomes a victim of traumatic events. As a victim, you will want to know exactly what happened to you, so that you may know exactly what you must do to recover and to get on with normal living. Understanding will help you to feel less alone with your secrets of abuse, and understanding will give you the power to overcome your painful feelings.

Psychological Trauma

Psychological trauma is the state of severe fright that we experience when we are confronted with a sudden, unexpected, potentially life-threatening event over which we have no control, and to which we are unable to respond effectively no matter how hard we try. Witnessing such an event can produce the same severe fright. Having your house destroyed in an earthquake, watching your father torture your pet dog, being left paralyzed from a street mugging, being raped by your parent are all examples of sudden, unexpected, potentially lethal situations over which we have no control.

Victims and witnesses are usually at first stunned, then highly frightened, and then angry. The body's and the mind's emergency response systems are activated. Heart, lungs, muscles work at top efficiency as do attention, concentration, and memory. This readiness ensures that victims will be able to cope with the threat of potential annihilation as best as can be. It is in this process that personal control, caring attachments to others, and the ability to make meaningful sense of life become disrupted.

Victims can be in acute crisis anywhere from a few hours to a few days. Fear and anger may be followed by confusion and withdrawal from others. Some victims have trouble sleeping, and many have recurrent memories of the event.

These are normal reactions to a terrifying stressful life event. Within a short time the fear is lessened, mastery is restored, time is spent with loved ones and friends, and some initial way of understanding what has happened begins to emerge. The psychological trauma has passed, and the victim is on the path to recovery. But there are exceptions.

Posttraumatic Stress Disorder

Such expected readjustment is not found in all victims, however, so let us examine this process in greater detail.

There are three domains of human functioning that contribute to good physical and mental health and a sense of well-being. They include reasonable mastery (the ability to shape the environment to meet our needs), caring attachments to other human beings (friends and relatives who care for us and provide emotional support), and a meaningful purpose in life (a reason to energize us to become involved in the world each day). Most of us go about our daily lives with these domains more or less intact and we never really think about them per se.

As noted above, traumatic events change all of this. When any of us is confronted with or witnesses actual or threatened death, serious injury, or a threat to one's physical integrity and at the same time experiences intense fear, horror, or helplessness (American Psychiatric Association, 1994; Everly and Lating, 1995; Flannery, 1999; van der Kolk, McFarlane, and Weisaeth, 1996), each of these domains associated with good health may be disrupted. By definition, we cannot control these traumatic events so reasonable mastery may be disrupted. We as victims may withdraw from our caring attachments as they may have been the assailants (incest), may have been victims also (a natural disaster that scatters a neighborhood), or because we want to withdraw from what is experienced as an unsafe world. Finally, the world may not seem safe, predictable, and worthy of our continuing to invest energy in it. Our sense of meaningful purpose becomes a causality of the traumatic incident. As if this were not enough to have to deal with, some traumatized persons may also develop characteristic symptoms that include unpleasant, heightened physiological arousal; intrusive recollections of the events; and a desire to avoid anything that is a reminder of the crisis (see below).

For many direct victims or witness victims, the disruptions in the three domains and the characteristic symptoms pass fairly quickly and normal life returns. For others, however, the disruptions and symptoms may linger for weeks, months, years, and even lifetimes, if they are left untreated. Victims in these latter circumstances may develop the medical condition known as posttraumatic stress disorder (PTSD), the stress and disordered functioning that continue, after the traumatic event itself is over.

If victims continue to experience distress during the first three months after the incident, they are experiencing *acute PTSD*. If the distress has continued into the fourth month or beyond, the victims are experiencing *chronic PTSD*. If the victims were seemingly not distressed at the time of the traumatizing event and kept on with their normal routines but then experienced the symptoms and disruptions for the first time six months or more after the event, the victims are experiencing *delayed onset PTSD*.

Let us turn now to some examples of PTSD. Acute PTSD may be found in victims of natural or man-made disasters, such as hurricanes or car and train accidents. These victims may experience loss of mastery, possible disruptions in their network of caring attachments, and may temporarily put aside their meaningful purpose in life to cope with the immediate crises at hand. These victims may experience intense fear and hypervigilance, may find themselves thinking about the crisis constantly, and trying to avoid unnecessary reminders of the painful situation.

Chronic PTSD, for its part, may be found in some victims of combat, physical abuse, and/or sexual abuse. Although some semblance of daily routines may be in place, the impact of these painful events remains past three months. Victims doubt the readiness of their basic coping skills, caring attachments are still avoided, and trusting others may have become a real concern. The world does not seem orderly and safe and something victims would want to be a part of. The sense of meaningful purpose is lost. Without PTSD treatment, victims may continue to experience these distressing events until death.

Delayed onset PTSD is still a medical and scientific mystery and I will return to this issue later in this chapter. For now, let me share an example. A nurse is assaulted by a young male patient in a hospital emergency room. Although the punch was painful, the nurse reports that she is fine and continues on her shift without any apparent distress. Ten months later when she is driving home from work, she comes across a motorcycle accident and provides CPR to the victim who is a young male. Suddenly, she becomes frightened and starts having memories of the emergency room fight. She is experiencing delayed onset PTSD.

We will discuss other examples as we go along, but let us review again the basic sequence of events as we begin. A person has intact functioning in the three basic domains and is leading a relatively normal life. Psychological trauma befalls the person and the person copes as best as he or she can. However, if normal functioning does not occur within a short period of time, the person may develop PTSD in either its acute (three months), chronic (four or more months), or delayed onset (first appearance after six months) state. The PTSD may include disruptions in the three domains as well as the characteristic symptoms.

We continue now with a more thorough examination of the symptoms that are frequently present in psychological trauma and PTSD.

The Symptoms of Posttraumatic Stress Disorder

Every medical and psychological problem gives us signals from our minds and our bodies that things are not right, that we are not in a normal state of health. The signs or symptoms of PTSD are listed in table 2. These symptoms may be present during the traumatic event itself as well as in any phase of the PTSD period. If you have PTSD, you may recognize some of these symptoms in your own life.

Physical Symptoms

The first grouping of symptoms has to do with physiological distress. Hypervigilance is a continuing state of alertness, even when there is no crisis. The person is so preoccupied with maintaining safety that he or she cannot relax. An exaggerated startle response refers to that of an individual who jumps at the slightest noise, even when there is no danger. Both hypervigilance and exaggerated startle response are common in untreated victims.

Sleep disturbances are also frequent, especially problems falling asleep and staying asleep. Victims may also experience difficulties in concentrating on matters at hand. They may find that they cannot recall things easily, or that they cannot shut off past memories, or both. Victims of violence sometimes have mood irritability, especially flashes of anger, when there is no apparent reason to be angry. Often they remain depressed.

TABLE 2

Symptoms of Posttraumatic Stress Disorder

Physical Symptoms	Hypervigilance
	Exaggerated Startle Response
	Difficult Sleeping
	Difficulty with Concentration or Memory
	Mood Irritability—especially Anger and Depression
Intrusive Symptoms	Recurring, Distressing Recollections (Thoughts, Memories, Dreams, Nightmares, Flashbacks)
	Physical or Psychological Distress at an Event That Symbolizes the Trauma
	Grief or Survivor Guilt
Avoidant Symptoms	Avoiding Specific Thoughts, Feelings, Activities or Situations
	Diminished Interest in Significant Activities
	Restricted Range of Emotions (Numbness)

Intrusive Symptoms

When a person encounters psychological trauma, the brain alternates between recalling the painful event to better understand what has happened (intrusions) and going to great lengths to forget it and the pain associated with it (avoidance). Each process has its own set of symptoms.

The intrusive symptoms deal with unwanted recollections of these past painful events. Thoughts, memories, dreams, nightmares, and flashbacks (a special form of memory that is called dissociation) are all common forms of recurring, daily, intrusive experiences of the traumatic events. For example, some victims are always dwelling on the event, others know that each night will bring a nightmare that reenacts the violence they have been through. Other victims depersonalize the event, and experience the traumatic episode as happening to someone else. It is as if they were in the audience of a theater watching the violence happen to them on stage. Still others have a memory problem called amnesia in which they cannot remember anything about certain events or periods of time in their lives. Events that symbolize past traumatic memories can produce painful physical and psychological distress on a moment's notice.

Grief and survivor guilt form the last type of intrusive symptoms. Victims often experience grief over the loss involved in the traumatic episode. The loss of one's house in a natural disaster, of one's virginity in incest, of one's physical integrity in battering are all sources of grief, as is the loss of loved ones in murder, combat, or sudden death. Grief is an intrusive type of symptom because it continuously reminds us of the loss. Survivor guilt is a special form of grief (Danieli, 1989). The victim experiences guilt that he or she has survived when others have not. Such grieving additionally offers the survivor a chance to continue the sense of community with the deceased and to honor their memory. Survivor guilt is often a way of fending off hopelessness.

Avoidant Symptoms

The third and final grouping of symptoms contains those in which the victim goes out of his or her way to avoid any specific thoughts, feelings, activities, or situations that could remind the victim of the traumatic event and bring an unwanted return of painful, intrusive recollections. Victims also avoid crowds, avoid friends, avoid promotions, avoid fun, and similar events that could make life enjoyable. Such avoidance leads to a diminished interest in significant and pleasurable activities and a restricted range of emotions.

All of these symptoms are unpleasant, with sleep disturbance, nightmares, and flashbacks considered to be the core symptoms of PTSD (Lavelle and Mollica, 1985). Since the symptoms in table 2 may also be signs of other medical problems, each victim should be evaluated by a physician

Delayed Onset of Symptoms

As we have noted, the symptomatic distress that occurs at the time of the traumatic event abates in some victims, and may not return until months, or in some cases, years later. Medicine and psychology are still attempting to understand this.

Sometimes victims encounter an event in their environment that reminds them of the past traumatic situation. A woman, battered as a child, may be inadvertently kicked by her husband when they are sleeping, and this event may remind her of her childhood abuse. A victim of street crime may be watching fireworks. The explosion of the fireworks may remind him of the gunshots at the time of the crime and revive vivid, intrusive memories that had been dormant.

In other victims, loss seems to be the major precipitant for the revival of intrusive memories. The loss can be the death of a loved one, the loss of a job, a geographical relocation. One of my colleagues, psychiatrist Bessel van der Kolk (1987), reported on a fifty-five-year-old female who was brought to the hospital from a supermarket where she was screaming in the aisles that everyone had to get out because of an intense, smoky fire. There was no fire in the store, but at age nineteen this woman had worked in a nightclub where a fire had resulted in severe loss of life. The woman testified at a fire commission hearing about the event one year later, and then promptly got on with her own life. For the next several years, she functioned very well in her business and personal life. When she was in her early fifties, the patient's father was diagnosed as having a terminal illness. It was the father who had gotten the patient her job in the nightclub. As the father's health failed, the daughter started having flashbacks and intrusive memories of the deadly fire forty-three years earlier. The loss became an avenue for the recall of symptoms many years later, symptoms which were then successfully treated.

The Experience of Symptoms

In this and the next two chapters I want to consider the seemingly ordinary behavior of a seemingly ordinary person from the perspective of PTSD so that we can think about the aftermath of traumatic events and how they may continue to have an impact on our daily lives, even without our realizing it. Let us begin with Joan's day's events first.

Joan is a twenty-seven-year-old administrative assistant in a small law firm. She has always been a shy and solitary person who found parties and large gatherings unpleasant, but with her methodical self-control she was able to accomplish much on her own. Periodic bouts with depression left her largely adrift of longer-term aspirations, but she somehow managed to keep going.

The law firm job seemed suited to her needs. It was small and manageable, there were not great numbers of people to deal with, and it would fill her hours until she could figure out something more useful to do with her life.

At eleven o'clock on this particular day, her boss and one of the firm's lawyers got into a huge argument over a particular case. The argument left Joan uncomfortable all day. She could not get her work organized, and by early afternoon she had developed a bad headache.

At home that evening, Joan left the lights on in rooms she was no longer in, left the tap dripping at the sink, and ended up burning the evening meal that she had been preparing.

Her husband, sensing her distress, commented to her that she seemed tense and preoccupied. Joan turned on him with fury, accused him of ordering her around, and threatened to leave him if he did not stop. At that point she barged out of the kitchen and spent the remainder of the evening in a room by herself. She seemed calm at bedtime.

How are we to understand this? Is this another example of a person having a bad day? In this case probably not, if we know that Joan has a past history of traumatic abuse. She repeatedly witnessed verbal and physical abuse between her parents. With this piece of information it is possible to understand the day's events from the point of view of PTSD.

When the argument occurred at the office, it symbolized for Joan the past angry arguments between her parents, and her body went into hypervigilance and the other forms of anxiety associated with the physical symptoms of PTSD. By supper time it is very likely that her inattentiveness to the lights and the meal was caused by intrusive memories or feelings of the bitter events between her parents many years earlier. She may even have been having flashbacks that she did not tell her husband about. When he came to her aid and tried to be supportive, she withdrew from him, and used the avoidant symptoms of PTSD for the remainder of the evening until her physical symptoms abated.

The stirring of symptoms like those presented here is not an infrequent event in the lives of victims of posttraumatic stress disorder, and such intermittent disruptions can be as frustrating and wearisome as symptoms that are continuously present, like hypervigilance or depression.

Dissociation, Learned Helplessness, and the Repetition Compulsion

With this basic understanding of psychological trauma and PTSD, we need to turn our attention to three additional psychological processes that are frequently core problems for victims of trauma. These include dissociation, learned helplessness, and the repetition compulsion.

Dissociation

Dissociation is a curious and, as yet, not fully understood process by which the brain puts a painful event on hold by splitting off the event from normal conscious awareness. It is a special form of an intrusive symptom.

In the normal process of consciousness we are aware of all that we are doing and we can remember each step of the task at hand. For example, most of us brush our teeth in the morning. We know where the toothbrush is, we know how to brush teeth, and we complete the task successfully. This particular episode of brushing our teeth becomes consolidated in our memory in that section or schema of our memory that stores tooth-brushing experiences. This newly formed memory is thus available to our waking brain the next morning. We have not forgotten the memory and we are able to retrieve it at will.

Conscious awareness and memory in the face of overwhelming traumatic events can differ in important ways from this normal process. What happens, for example, to the combat soldier in the heat of battle who must walk among the body parts of his fallen comrades even as he pursues the enemy? One thing that can happen is that the brain may enter the process of dissociation. The brain places the traumatic event of seeing the body parts out of immediate consciousness by creating a separate consciousness for that event. This dissociated consciousness has its own feelings, its own thoughts, its own physical body movements. It is stored in the brain as a unitary memory (since the victim has no similar past experiences), and is quickly beyond the conscious control of the victim. In some cases the victim may "forget" or develop an amnesia for the dissociated experience, but it may return unbidden in subsequent experiences with any event that symbolizes the trauma.

To continue with our example, the combat veteran returns home. He knows that his war experiences were frightening, but he remembers them only in generalities. One day he is walking down Main Street with the intention of paying his utility bill. A bus backfires. That sound symbolizes his war experiences, the dissociated memory returns, and our veteran dives behind a parked car, starts hiding from the "enemy," and vividly recalls the memories of his fallen comrades. All of this has happened on Main Street. This return of the dissociated memory is an intrusive symptom called a flashback and is common in victims. For example, rape victims may have such flashbacks when they make love to their spouse; battered individuals might have such an experience if a friend waves vigorously at them; a car accident victim might have flashbacks at the sight of blood.

Why does the brain do this? The process is far from understood, but it seems to be the brain's way of enhancing survival in the face of life-threatening traumatic events. When confronted with such an event, we need to focus our full attention on staying alive. This is not the time for us to be flooded with grief, incapacitated with anxiety, or repulsed by what is happening to us. The brain helps us out by taking the aspects of the terrible event and dissociating them, so that we can distance ourselves from them psychologically. With the catastrophic event thus somewhat removed and contained, we can focus on survival. The dissociated event is stored in memory until the life-threatening event has passed.

As time passes, the dissociated event intrudes upon consciousness and the victim assimilates some of it and processes it before it is again dissociated by the passage of time or by active efforts on the part of the victim to avoid it. Gradually, over time, this process of intrusion-avoidance allows the brain to process the information and to fully incorporate the dissociated component that could not be dealt with at the time of crisis. This appears to be nature's way of helping us to survive and to return our lives to normal (Horowitz, 1986; Litz and Keane, 1989; McNally, 2003).

Signs of Dissociation. All of this sounds strange and puzzling, yet most of us have had some form of normal dissociation. Daydreaming, "tuning out" a conversation, and highway hypnosis are all common forms of mild, normal dissociation. In each case, the event split us off from more normal conscious awareness.

If you were a victim, how might you know if you had the more serious forms of dissociation? Psychiatrist Frank Putnam and his colleagues (Putnam, 1989) have given us some examples. A victim may find herself in clothing that she does not remember putting on. Similarly, another victim might find writing or drawings by his own hand, but not remember doing them. Some victims may look into a mirror but not recognize themselves, or be in a familiar place that still feels strange. Other victims may find themselves accused of lying when they do not believe they have lied. All of these are examples of what may be the more severe forms of dissociation, and all victims are more prone to dissociative experiences if they are intoxicated, ill, fatigued, depressed, or very young when the traumatic event happens.

While dissociations can be successfully treated, untreated experiences of dissociation persist. A recent study of the Dutch Resistance Fighters of World War II reported finding flashbacks and similar intrusive memory problems in soldiers now in their sixties and seventies. Old soldiers may never die, but neither do their untreated dissociative experiences.

If you want to read further, psychiatrists Bennett Braun (1986) and Frank Putnam (1989) have written at length about this unique aspect of human functioning.

Learned Helplessness

We noted that victims of traumatic events are unable to do anything to solve the crisis at hand. Some victims further compound the problem of loss control at the time of the crisis by making a false assumption: they assume that because they could do nothing about the traumatic event they can do nothing about other life events either. They lose the capacity to appreciate the connection between their actions and shaping the world to meet their needs. They have incorrectly generalized from one situation to all situations. They stop trying to master the environment, and they learn to be helpless. Such helplessness is rarely reported in victims with one traumatic event in their lives, but helplessness is often found in victims of repeated abuse.

Helpless victims believe that they exercise *no reasonable control or mastery* over their lives. Such persons assume that any efforts on their part to shape the environment to meet their needs will fail. Such persons may stop trying to cope altogether, and some of them rely heavily on others to solve their problems for them.

The second characteristic of helplessness is a state of passivity or noninvolvement in the world. Helpless people have *no personal commitment* to anything or anyone in life that would make life meaningful. The helpless person remains uninvolved with career, family, raising children, community projects, and the like.

A third characteristic of helplessness is that of *disrupted daily routines*. Persons with helplessness have problems planning their daily and weekly routines. They do not eat regular meals; they do not sleep at regular intervals. Work may become sporadic. Interest in sexuality, recreation, and the other normal pleasures of life may be similarly disrupted.

Finally, helpless persons are often *socially isolated*. They avoid being with others at work, at home, and in the community. The nature of the painful events has left them apprehensive, if not outright distrustful of others, and they withdraw into themselves.

The end result of these four characteristics is a painful state of depression that can last for years. As long as the person remains helpless, there is little reason to expect that the depression will lessen because the very characteristics of helplessness keep the person from engaging in ways that will break this cycle.

Perhaps the most interesting aspect of helplessness is that not all trauma victims go on to become helpless. Our understanding of this is not yet complete, but we do know that men and women who have excessive needs for personal control, and who hold themselves accountable when others would not, are at increased risk for developing helplessness. Additionally, those victims who see their traumatic events as one more instance of a long-standing pattern of unhappy life events, and who assume that such events will be of long duration are more likely to develop learned helplessness.

Helplessness That Is Not Helpless. Equally important to note is the fact that not all behavior that appears to be helpless is in fact learned helplessness (Flannery and Harvey, 1991).

Some victims who appear helpless are blocked in their efforts to gain mastery by others. For example, a combat victim seeks help at a military hospital and is told his disabilities are not service-connected. An adult incest survivor discloses her abuse only to find herself cut off from the family. A battered spouse cannot obtain an effective restraining order.

Some victims appear helpless and passive in the face of ongoing traumatic events because in fact they lack the skills to solve the problem or to escape. What to do? When to leave? Whom to trust? What to say? These are complex skills. Physically or sexually abused young children may not be able to formulate such questions. Elderly and handicapped victims may lack the physical strength to act.

Other victims appear helpless because they believe that they do not deserve better. Self-esteem remains poor even when the individual has the skills to cope more effectively. Youngsters growing up in alcoholic homes are frequently criticized relentlessly and are sometimes physically or sexually abused. As adults, these victims may assume that continued victimization is justifiable punishment for their presumed inadequacies.

Still other victims appear helpless as a strategy for survival. By appearing to be passive, they hope to minimize or contain the abuse and the chances for possible injury. When other possibilities exist, their coping strategy is abandoned in favor of direct escape.

Lastly, some victims of crime or of physical or sexual abuse may appear helpless and accept the abuse as a way of protecting others who seem more weak or fragile. Such altruism may motivate a wife to absorb the violence to divert it from the children.

A daughter may endure repeated incestuous abuse in hopes of sparing a younger sibling. Such victims may be highly skilled but be unwilling to use these skills lest someone else be hurt.

These cases of apparent helplessness are really active attempts to solve problems. There are certainly better solutions as we shall learn, but such behavior is not learned helplessness either.

We can see from the preceding discussion that the process of learned helplessness in victims of chronic abuse is very complex. In those instances when victims make the false assumption that they can do nothing about life's problems, they usually assume that no one else can help them either. Such a view includes loved ones and counselors too, so that in helping such victims to recover fully, the problem of the learned helplessness must be resolved first before the traumatic event itself can be addressed. We shall see how to do this later on.

The Repetition Compulsion

One would think, with the experience of a traumatic event followed by the phases of PTSD and its painful symptoms, that victims would withdraw and take time to rest and heal their wounds. Often just the reverse is true: some victims actually seek out situations that remind them of the past traumatic events. They seek to repeat or reenact the trauma.

Some combat veterans provoke fights in bars on weekends and usually lose. Some rape victims become prostitutes. Some battered children become self-mutilators or later, when they are adults, batter their spouses. Children of alcoholics may become addicted to alcohol or drugs.

The question is why fragile, decent people who have already been overwhelmed by traumatic abuse would seek to repeat it.

From a psychological point of view, the repetition compulsion might be an attempt to regain the reasonable mastery that was disrupted by the original traumatic event. Such attempts would provide ways to master the situation anew so that meaningful sense can be made of the event. Thus, victims would learn what is the best way to cope should such a horrible event befall them again.

But why do the victims actually re-create it in their lives? Why not just think it through at some later time? Why return to the lion's den and deliberately expose oneself to possible further danger and pain? A current and more likely explanation is that these victims have limited ability to see and utilize better solutions to their problems.

Finally, some victims engage in repetitive addictive behaviors that foster the enhancement of chemicals in the brain, known as endorphins, that make us feel relaxed and calm. Since their presence in the brain produces relief, they are, in effect, a form of self-medication for the painful distress that results from untreated PTSD.

Recent medical evidence suggests that behaviors, such as sex addictions, binge eating, cocaine and crack addictions, fist fights, self-mutilation (e.g., repetitive wrist cutting), reckless behavior, and accident-proneness all appear to produce this endorphin release.

Thus, it is reasonable to assume that some persons with addictive behaviors and a past abuse history may be using their addictions as a way of self-medicating the distress of the traumatic abuse by creating an endorphin release.

This would mean that such addicted persons have two medical problems in need of treatment: the addictive behavior and the untreated past abuse. If the addicted victim treats only the PTSD, the addictive behavior, which in many circumstances has itself become physiologically rooted, will continue, and may complicate recovery from the traumatic event. Likewise, if the addicted victim treats only the abuse, the physiological distress from the symptoms of untreated trauma will leave the victim in physical distress and possibly lead to further addictive behavior—continuing attempts to self-medicate that distress again. Further, untreated traumatic distress is marked by irritability, anxiety, anger, and interpersonal hostility symptoms similar to many of the signs of a "dry drunk." While some of these signs may be the result of long years of alcohol abuse itself, it may also be that some "dry drunks" are experiencing untreated PTSD distress. For all of these reasons, it is important for addicted victims to treat both the addictive abuse and the PTSD aftermath.

*   *   *

Our purpose in this chapter has been to gain some understanding of the nature of psychological trauma and its symptoms, and to see if these matters bear any relevance in your own lives. Do you have any of the PTSD symptoms? Have you had dissociative experiences? Do you feel helpless and depressed? Are you struggling with an addiction? Does any past event in your life now appear to have been a traumatic event as we have outlined it here?

It is often helpful to discuss such matters with someone you trust: a loved one, a friend you can trust, your physician, your counselor, a member of the clergy. This caring person will help you evaluate the issues raised in this chapter.

There is relief from the suffering and pain of psychological trauma. Many victims have learned to overcome their symptoms, and to master helplessness, the repetition compulsion, and addictive behavior. You can do these things too. In raising your awareness of trauma and its possible impact on your life, you have already started your recovery process.

2

Am I Losing My Mind?
The Psychology of
Posttraumatic Stress Disorder

Tis all in pieces, all coherence gone.
— John Donne

You are a child of the universe.
You have a right to be here.
— Max Ehrmann

The locking on of brakes, the screech of tires, the smell of burning rubber, the crush of metal on metal, the shattering of glass. And then silence.

"He's blue. He's turning blue. Someone help him. He's dying."

"It's okay, lady," said the police officer who had responded to the call. "The EMTs are here; we'll have you safe in no time."

"He's blue. He's dying. Please help him."

"I've check the other driver. The witnesses say he was speeding when he ran the light. He's badly shaken up, but he's not blue," replied the officer in his attempts to calm the obviously frightened and severely agitated female driver.

With Paula extricated from what had been a motor vehicle, the EMTs cleansed the abrasions, set the splints, and sped quickly toward nearby St. Luke's emergency room. Their patient was in shock—both physical and psychological.

When William, her husband of some twenty-five years, arrived at the hospital, the physicians asked him if he could make any sense of his wife's reference to the other driver's turning blue. Bill could not. Paula had never had any accidents that he knew of in all of their years of marriage. He did remember that first summer that they were married he had come home on a hot and humid late August afternoon to find his wife with the stove on high. She told him she was cold. He had seen this phenomenon every year. Paula was always irritable before Labor Day. But, no, he could relate none of this to turning blue.

Paula remember. She recalled that sunny day on the county road. She and her parents were returning from summer vacation at the lake. School was to begin next week. Seven-year-olds can have a lot of fun at the lake in the summer, and she and her little friend Henry Collins had gone swimming every day. Even when it rained.

Her parents were first to notice the accident. It was the Collins's car, and Mr. and Mrs. Collins were trapped inside. One of the Collinses had severely hit the now bloodied windshield. They heard the voice of Paula near the ground in front of the car. "Mommy, mommy, he's blue. Henry's turning blue."

The hushed and busied silence of the emergency room did not stop her tears nor soften the sting of grief. She had never thought about her little friend's death since the day it had happened. She lay quietly and listened to her heart beat. The memories were vivid and far more painful than her broken bones.

*   *   *

There can be no doubt that the effects of posttraumatic stress disorder are painful and frightening at any age. Paula is experiencing the delayed onset of intrusive memories of a long-forgotten episode brought to mind by her own car accident that day. As she lies there on the hospital gurney, she is having vivid flashbacks. Her mind is now remembering the sad events of that Labor Day weekend so many years before, the events that her body had remembered in the form of mood irritability and coldness (of death) each and every year since the day that it had happened. Her then young and growing sense of mastery, her links to her playmate Henry and his family, and her ability to make some meaningful sense of what had happened were profoundly dislodged at the time of the accident, and were locked from her awareness in a dissociated state. The severe life stress of psychological trauma can clearly alter our normal psychological processes for coping in very fundamental ways.

In *Becoming Stress-Resistant,* I wrote in detail about life's effective problem-solvers, whom I refer to as Stress-Resistant Persons (Flannery, 2003). Our understanding of these men and women is based on research that I directed for twelve years on twelve hundred men and women. My colleagues and I found that stress-resistant persons exercised reasonable mastery, developed caring attachments to others, and were committed to some goal in life that made their lives meaningful. By utilizing these three general sets of skills in dealing with the many stressful life events that each of us face every day, stress-resistant persons were able to maintain good physical and mental health as well as enjoy a sense of well-being, as we noted earlier.

As we have seen, mastery, attachment, and meaning are the very domains that are disrupted in psychological trauma and posttraumatic stress disorder. Subsequent clinical findings have taught us that stress-resistant persons or those with the various skills of stress-resistant persons are better prepared to cope with the severe life-stress of a traumatic event. Such skills can buffer the potential negative effect of the traumatic event itself as well as hasten the recovery process.

In this chapter, we shall study both these adaptive ways of coping with traumatic stress and the fundamental ways such events can alter this basic process. By understanding clearly what is helpful and how things go awry, we shall be better able to learn or to restore the helpful skills of stress-resistance that seem so beneficial for healing and recovery. These skills of stress-resistance can be readily learned, and every victim will want to use them in his or her own life.

As you read these pages on the psychology of posttraumatic stress disorder, make a mental list of which of the skills of stress-resistant persons you may have and those that you may need to learn. You will also want to pay close attention to how things become disrupted, and to note any of these less-than-adaptive ways of coping that you may be currently using. Such knowledge will be helpful later on in planning your specific steps to recovery.

Normal Psychological Adaptation:
Stress-Resistant Persons

Table 1 lists the six characteristics that stress-resistant persons employ to attain mastery, attachment, and meaning. From this list, we can see that they utilize personal control, basic life-style choices, and humor as strategies for reasonable mastery. Task involvement, social support, and the basic religious/ethical concern for the welfare of others are used to develop caring attachments as well as to provide for a meaningful purpose in life. Here is what stress-resistant persons have to teach us.

Reasonable Mastery

The world is complex and often beyond our full control, but the stress-resistant skills of reasonable mastery, sensible life-style choices, and a sense of humor (table 1) can help immeasurably to develop an effective sense of personal control or general mastery.

Personal Control.   Effective problem-solvers use a five-point process in their efforts to attain some sense of mastery. First, they clearly identify the specific problem to be solved, gather information to solve the problem at hand, reflect on possible strategies for resolution, develop and prioritize several strategies to solve the problem, actively implement a possible solution, and lastly, evaluate the proposed solution to see if it worked.

TABLE 1

Stress-Resistant Persons:

---

1. *Take Personal Control*
2. *Are Task Involved*
3. *Make Wise Life-Style Choices* —*Few Diet Stimulants*
   —*Aerobic Exercise*
   —*Relaxation Exercises*
4. *Seek Social Support*
5. *Have a Sense of Humor*
6. *Espouse Religious/Ethical Value of Concern for Others*

Not all problems are solvable. Some relationships do not work out, some promotions are not based on merit, some goals are beyond our income, and so forth. Good problem-solvers know when to accept as a given what cannot be changed, and this attitude of acceptance frees them to direct their energies toward more beneficial goals.

Life-Style Choices. Stress-resistant persons maintain reasonable mastery by keeping their minds and bodies fine-tuned. They avoid the dietary stimulants of caffeine and nicotine, participate in a regular program of aerobic exercise each week, and spend time engaged in some form of relaxation each day. The aerobics can be as brief as three twenty-minute periods over a seven-day time span, and the relaxation period may be as short as fifteen minutes each day.

Humor. People who cope effectively with life have a sense of humor or chum around with those who do. Laughter reduces the physiology of stress and helps us to keep things in perspective. Traumatic events are not funny, but being able to laugh at any of the other paradoxes in life is helpful medicine.

Caring Attachments

Caring attachments are the beneficial links we have to others. These links provide important health benefits to both parties. Our cardiovascular system, our immune system that fights certain kinds of infectious diseases, and our endogenous opioid system, which produces endorphins, also work more effectively when we are in caring relationships that include task involvement, social supports, and concern for the welfare of others (table 1).

Task Involvement. Everyone needs a reason to live. There must be something in the lives of all of us that we are personally committed to seeing through to completion because it is existentially important for us to do so. Improving one's career, raising one's children, serving on a community project are all examples of possible goals.

It does not mean that completing these goals will always be easy, nor that we always attain the goal. Such goals, however, motivate us to become involved with life, and guide our day-to-day interactions in fundamental ways with others.

Social Support. Effective problem-solvers are not social isolates. They realize the psychological benefits that can come about in friendships.

Caring attachments can provide us with emotional support when good things happen to us and when sorrow is our lot in life. Such support helps us to share events, minimizes our sense of being alone, and helps us to go on. Such relationships can also provide us with information about how to solve problems and when to stop such efforts. Caring others can also provide us with various types of assistance in the form of money, material goods, or political favors, and they can provide us with companionship through life. Sharing the human journey with others helps to reduce our sense of helplessness and aloneness, and provides meaning for the events of daily living (Lynch, 2000).

Espouse Religious/Ethical Value of Concern for Others

A third way adaptive problem-solvers develop caring attachments is in following the basic psychological tenet in all the great religions of the world: the Golden Rule. Do unto others as you would have them do unto you. This concern for the welfare of others, when put into specific actions, builds links to others and helps us to find meaning in life. Assisting a family member, coming to the aid of a neighbor, helping a stranger in time of crisis may all lead to new or strengthened supportive attachments, with all of the physical and psychological health benefits that we have noted.

Purposeful Meaning in Life

Stress-resistant men and women have taught us that each of needs a purposeful reason to live. Such a reason provides meaning in life and the motivation to go forward, particularly in life's darkest moments when we want to give up on everything, including ourselves.

Sense of Coherence. Sociologist Aaron Antonovsky (1979) has written about the importance of each of us having a meaningful sense of coherence about life. Our sense of coherence is how we view the world, and it provides us with a perspective on human events. It has three components. First, one's sense of coherence must provide a sense of manageability, a belief that we can exercise some reasonable control over our environment and shape it so that we can attain, to some degree, the goals that we seek in life. Second, one's sense of coherence must help to make the world comprehensible in that it must help us to understand and to predict the events in our lives with some reasonable degree of accuracy. Third, the sense of coherence must provide some specific meaning for each of our lives individually so that we believe life to be worthy of our investment.

Taken together, the components of the sense of coherence help us to keep life events in perspective. In addition, nature tries to help us out with this process with three mild illusions or exaggerated beliefs about ourselves. These occur as a normal part of growth (Goleman, 1985; Taylor, 1986), and help to buffer us from the stress of life.

The first illusion is one of self-enhancement. Most of us really evaluate ourselves in a much better light than objective fact would warrant. We feel that we are more attractive, more talented, more skilled than is the actual fact. The second illusion is a somewhat exaggerated belief in our sense of personal control. We feel that we are better able to cope in some situations than we are actually able to do. The third mild exaggeration is our unrealistic optimism about the future. Most of us really expect better outcomes in life than life may actually provide.

These mild illusions in normal people do not lead to arrogance; rather they appear to be Nature's way of helping us to keep our chin up when times are bad, and these beliefs help to strengthen our sense of coherence. Without these beliefs life is more difficult.

Concern for Others.   There are many ways each of us may find meaning in life. Our culture encourages us to seek material goods and to attain all of our personal strivings. Power, recognition, money, material goods, political influence are all presented as ways to attain meaning in life.

It is curious then that stress-resistant persons, those who have better health and well-being, should seek out caring attachments to others as their source of purpose in life. For a purposeful sense to life they choose task involvement, social support, and a concern for the welfare of others (table 1). These three characteristics are concerned with others—living for others, giving to others, brightening the world of others.

Why is it that stress-resistant people do not seek power, fame, money and so forth as their primary meaningful goal in life? The late sociologist Ernest Becker (1973) has written an interesting book that may help us understand their choice of others. He has noted that the human person is half animal and half spirit. The half of the person that is animal knows that it must die at some point in time. The half that is spiritual seeks to transcend his or her own mortality and human finiteness, to find some way to live on after death. Power, money, recognition, material goods, and the like all are finite like the animal half of the person, and, like the animal half, all end in death. As such, these goals cannot provide for us any way to transcend death and to live on in some way after our own physical demise. But caring for others can.

We can care for others in many ways that will allow us to leave our legacy after death. We may have children of our own, enter the helping professions, perform various acts of personal charity, make artistic and creative contributions, or endow foundations for noble work. There are as many ways to accomplish this as there are creative, caring persons. What is common to each of these ways is that the individual's meaning in life is rooted in concern for the welfare of others in some transcendent manner, and the medical evidence reports that a transcendent purpose in life results in better health and well-being. Such transcendence provides meaning in life even in the face of suffering, disability, difficult loss, evil by others, and even our own deaths.

*   *   *

Mastery, attachment, and transcendent meaning thus form the basic psychological components of sound physical and mental health. These skills can be very helpful to victims in recovering from the aftermath of traumatic events, as we have noted, and we shall return to them in the last section of the book. For the present, however, let us see how these helpful skills are disrupted in the face of traumatic violence.

Psychological Alterations in Traumatic Events

When the rape, battering, combat, or other traumatic events occur, the victim's normal psychological construction of the world is torn apart, as we have noted. Mastery is lost, support networks are uprooted, and the victim's sense of an orderly world is replaced with confusion.

Although the impact of any traumatic event varies widely from individual to individual, there remain some common ways in which victims are generally disabled.

Faulty Mastery

Just as personal control is associated with better health and well-being, loss of control leaves an individual in high stress. Victims often feel acutely vulnerable, confused about what to do next, and fearful of repeated episodes of violence.

Being master of one's destiny is a basic human desire, and there are some common but not particularly helpful ways that victims seek to restore this control. One such common method is to revert to a state of super-control. In such cases, victims essentially say to themselves: "I will never be this vulnerable again," "I will protect myself from all future unforeseen events," "I will always be in charge." Such persons then engage in methodical efforts to control most everything. Potential difficulties like financial matters become controlled; pleasurable events and relaxation become controlled; one's interactions with others become controlled. Such methodical overcontrol is not an adaptive way to respond. It is based on the false assumption that any person can control all life events, including future traumatic events. Such overcontrol usually leaves victims vigilant, frustrated, and unhappy.

A second common method to restore this sense of personal control is to blame one's self for what has happened. While this might at first seem paradoxical and self-defeating, it gives victims the illusion of control because it implies that, if victims had behaved differently, the event would not have happened or might have been better managed. Statements such as, "If I wasn't wearing my tennis shorts, I wouldn't have been raped," "If I hadn't crossed the street at dusk, I wouldn't have been mugged," "If only I hadn't sneezed, my father wouldn't have gone drinking," all imply that the victims would have had control of the situation if only they had been more in charge.

Self-blame as a long-term solution to regaining control is not helpful because it may leave the victim with hypervigilance, poor self-esteem, pervasive shame, and continuous depression. As a short-term coping strategy for victims, however, self-blame may actually be helpful in the first days of the crisis. Self-blame can mobilize the victim to focus on what needs to be done to restore personal order, whereas blaming others could result in helplessness or in the alienation of other potentially caring persons who could be helpful. Victims are better served if they think through the issue of retribution with the appropriate police, legal, and counseling authorities, and avoid indiscriminately blaming others.

The two remaining strategies for dealing with self-control are also not very effective. The first strategy is to make the false assumption that we have noted: to assume that nothing can be done, to give up altogether, and to remain helpless in the face of events. Such a stance will very quickly result in depression. The other ineffective strategy is to seek to control the biochemical distress of the traumatic event by developing a pattern of addictive behavior.

All of these strategies fail because the victim is kept from finding a more adaptive resolution to the traumatic event.

Inadequate Caring Attachments

Whenever a person becomes a victim of violence at the hands of another person, self-preservation as well as basic common sense lead the individual to withdraw, and to recoup energies for another day. No one who has been raped, assaulted, fought in war, or been otherwise attacked wants to run the risk of such painful events happening again. Unfortunately, many victims overgeneralize from one painful encounter with one harmful person, and tend to withdraw from everyone. They incorrectly assume that almost no one can be trusted. While this initial mistrust passes in many victims in the first few days after a traumatic event, for other victims, especially those in the chronic PTSD phase, this general mistrust of other persons may become a fixed and incorrect way of viewing the human family. The whole human family is not the enemy, even if it feels that way at times.

To be sure, there are some types of human relationships that are harmful and not caring. All of us may inadvertently cause each other sadness and pain because our own anger, frustration, or anxiety leads us to speak or behave in ways that are hurtful to others. These types of events happen periodically among people who care for each other. It is part of being human. When researchers identify harmful interpersonal relationships, however, they are referring to specific patterns of behavior that are continuous in the relationship and most often willfully inflicted. Five general types of potentially harmful situations have thus far been identified.

The first is basic value conflicts. Each of us is guided by basic values about the meaning of life, what it means to be a person, what it means to be a responsible member of society, what it means to love another person. None of us agrees on everything. That would make life dull. Notwithstanding, the greater the number of fundamental differences in these basic values, the greater the likelihood for a disrupted relationship. If you believe in treating others with basic respect, and your spouse sees nothing wrong in hitting people, you have a fundamental value conflict.

A second potentially harmful strategy may result when one party in a relationship is always demanding that things be done one way, or is always demanding to be the center of attention. This is called emotional demandingness. In the face of repeated and insistent requests, reasonable men and women will withdraw. Victims have their own legitimate needs, but if they are emotionally demanding, other persons who are potential resources for recovery will dissolve the relationships. If victims are in relationships with emotionally demanding others, the victims' needs will never be met.

Emotional over-involvement is a third form of a potentially harmful relationship. Over-involvement entails the use of excessive control over another person. Excessive overprotection, intrusions into other peoples' lives and excessive self-sacrifice, leaving others feeling guilty, are all associated with negative health consequences. As with emotional demandingness, victims can neither afford to be over-involved in the lives of others nor allow that to happen to themselves.

Interpersonal skill deficiency is a fourth potential form of a harmful relationship. Adults need to know how to share, how to argue fairly in a marriage, how to manage financial resources, how to raise children, how to be responsible employees at the worksite. When people are lacking such skills, the whole task of getting on with life by cooperating with others is jeopardized. Victims who must deal with such persons will find recovery all the more complicated because basic resources for problem-solving are absent when they are most needed.

Finally, relationships marked by physical, sexual, extreme verbal abuse, and/or either physical or emotional neglect are by definition traumatic and harmful events that impair health. Many people lead difficult and lonely lives; most of them do not abuse other people for any reason. There is no acceptable reason for remaining in an abusive relationship without seeking help to resolve it.

Victims will want to avoid the five types of relationships noted above, but it is equally self-defeating for victims to avoid the whole human family. Such an avoidance of others leads to the loss of the physiological and psychological benefits that we noted earlier.

In writing about the importance of maintaining links to others in the midst of traumatic events, I realize the understandable fear and reluctance of some victims reading these words to reach out and trust. Trusting others, especially after painful violence, need not be a blind leap into the unknown. Trusting others is not an all-or-nothing approach. There are clear steps that take much of the guesswork out of whether another person can be trusted or not, and I shall outline these basic principles in the section on treatment and recovery.

Loss of Meaningful Purpose

Just as mastery and caring attachments are dislodged by traumatic events, so the victim's purposeful sense in life is equally uprooted in the aftermath of violence. The impact of the violence on our reason to live may manifest itself in many ways.

Shattered Assumptions.   Just as Dr. Antonovsky (1979) has described the adaptive sense of coherence, social psychologist Ronnie Janoff-Bulman (1985) has studied how this sense of coherence is torn apart by traumatic events.

When a child is raised in a healthy home, the child learns important lessons that it takes with it through life. The child learns by interacting with its parents that there are caring people in the world, that the child is worthy of such care, and that interactions with others can have benevolent and helpful outcomes.

When the trauma strikes, these assumptions about a kind world are crushed. The first assumption to be lost is the victim's belief in his or her personal safety and invulnerability. The benevolence of the parent-child interaction is replaced by the reality of harshness and evil that exists in the world in many ways. The sense of protection (i.e., "It happens to others but it can't happen to me.") is dissolved immediately. Victims never again regain the sense of innocence that was theirs before the traumatic event occurred.

The second assumption to be challenged is the victim's positive view of himself or herself as having some reasonable mastery over the environment. For anyone who has been a victim, it is clear how fragile and tenuous human life is, how really frail we all are in the face of life's onslaughts. We are, in the writer Leo Therese's telling phrase, vessels of clay, and trauma molds our understanding, pain, and suffering.

The third assumption to be torn apart is the victim's belief that the world is a just, meaningful, and orderly place to be. Trauma brings us face to face with injustice, human meanness, and often chaos as we attempt to respond to natural disasters or interpersonal violence. Victims learn in a very direct way that life is neither fair nor understanding.

Our normal mild illusions of enhanced self-esteem, an exaggerated belief in personal control, and high optimism about the future (Taylor, 1986) flounder in the face of a traumatic event, as does our sense of coherence in a meaningful world (Antonovsky, 1979).

Negative Self-Evaluations. The loss of meaning in traumatic events can lead to further complications in the ways victims respond to such matters. In the absence of a meaningful way to make sense of what has happened, victims may learn to make incorrect negative self-statements or have specific psychological conflicts having to do with the meaning of such events.

Panic. Psychologist David Clark (Hawton et al., 1989) has provided us with one of the more up-to-date ways of understanding why some trauma victims develop panic attacks after violent events. It is not clear to researchers whether all panic attacks are learned or whether some types are medical illnesses. This question

FIGURE 1

Learning to Panic:

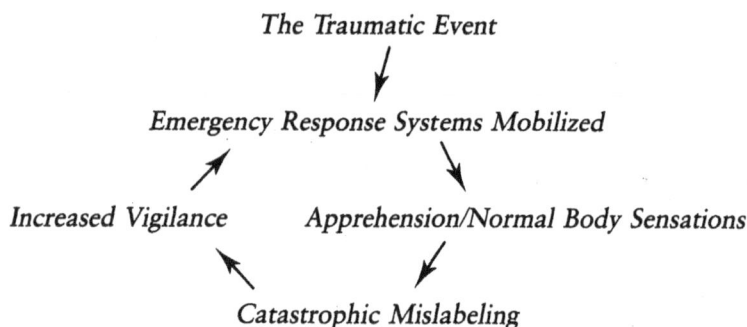

___

The Traumatic Event

↓

Emergency Response Systems Mobilized

↗   ↘

Increased Vigilance   Apprehension/Normal Body Sensations

↖   ↙

Catastrophic Mislabeling

___

awaits further research, but here is how Dr. Clark would explain how some cases of panic are learned (figure 1).

When violence occurs, the victim's body and brain emergency mobilization responses are activated as the heart beats faster, breathing is more labored, muscles tighten, and so forth. Some victims pay much attention to these bodily sensations, and label these normal bodily preparations for coping as catastrophic. The traumatic event is frightening to begin with, but some victims, in their attempts to make meaning or sense out of what is happening to them, compound this problem with a mislabeling of the body's basic response to severe stress. Victims make catastrophic statements like, "I will collapse," "I will faint," "I am going crazy," "I am having a heart attack," " I am dying."

Unfortunately, when the traumatic event is over and victims are again getting on with their lives, some victims do not correct the faulty mislabeling of the bodily sensations. Hence, anytime a muscle tightens, or breathing is labored, or the heart rate increases—as can happen in aerobics, sexual relations, or many other pleasurable activities as well as many lesser stressful life events—some

victims continue to misinterpret these bodily sensations, continue to make faulty catastrophic mislabeling, and end up inducing a sense of panic, even when there is no true cause for alarm. Moreover, victims begin to fear the onset of additional panic attacks, and begin to continuously scan the environment for possible threats. As such persons go about their daily routine in this heightened vigilance, ambiguous but innocent information from their bodies are falsely labeled as threats. The cycle for continuing panic attacks is set.

Negative Self-Statements. Continuous, and incorrect, negative self-evaluations can come about from faulty thinking patterns and from false assumptions about the self, but they do provide some illusion of control. While there is some evidence that some worriers are born that way, negative thinking is often learned, especially after traumatic events. Psychiatrist Aaron Beck (1975) has spent many years studying depression, and has identified some common faulty thinking patterns. Some persons overgeneralize from one event that goes badly to the conclusion that all events will go badly. Some persons develop the habit of discounting the many positive things they do right when they have failed at some one task. Others magnify their own imperfections, or assume an all-or-nothing mind-set in evaluating their capacity to cope.

Psychologist Albert Ellis (1963) has shown us the power of overbearing assumptions that can result in undue increases in anxiety and depression. Statements like "I am worthless when I act foolishly or badly," "I am bad and unloveable when I am rejected," "People *must* treat me fairly," and "It is horrible when major things do not go my way," are common examples of faulty assumptions found in some victims.

Neither form of negative self-evaluation is particularly helpful in making any sensible meaning of traumatic events. The victim had no control over the trauma, and should not spend a lifetime feeling guilty or ashamed.

Psychological Conflicts. As victims attempt to understand what has happened to them, some develop certain fears or worries that relate to their understanding of what has happened. These worries have remarkably similar themes regardless of the specific type of violence, and psychiatrist Mardi Horowitz (1986) has listed some of these more common worries or conflicts.

Some victims have a persistent fear that a similar event will happen again; others experience shame at having felt helpless when the traumatic event occurred. Some have continuing rage at the person who caused the violence, or the fear that they may themselves become violent should they encounter the assailant. Still others feel guilty that they harbor such retaliatory wishes.

As with panic and negative self-evaluations, these psychological conflicts are involved in attempts to make sense of the violence and to provide the illusion of control.

Grief. Every act of violence entails a loss, since something is taken from the victim. In rape, it may be the loss of virginity; in battering, the sense of personal physical safety. In combat, the innocence of youth may be crushed; in family alcoholism, trust in adults in often destroyed. Each loss represents the loss of some basic aspect of the victim's personhood, and such losses produce in the victim the same feelings of pain, anger, and sadness that a person might experience at the death of a loved one.

This state is called grief. It occurs when victims are separated from something important to them, and it has many signs. It may involve marked sighing; lack of energy; subjective distress, loss of sleep and appetite; feelings of hostility, depression, and guilt; and loss of interest in the world about us. Often a grieving person may be preoccupied with the deceased or with thoughts about his or her own death (Lindemann, 1944). Similar feelings are found in victims who sustain the types of important traumatic losses that we just listed.

One way of resolving grief has particular stages that all of us may go through (Kübler-Ross, 1969). *Denial* is the first stage, when the person refuses to believe that the event has happened at all or is as severe as it is. The second stage involves *anger.* The person now realizes the event has happened, but is angry at the limitations and lost dreams that such an event implies. *Bargaining* is third, as the person tries to barter with God or family and friends to mitigate the true impact of the event. This state is followed by *depression* as the person realizes the full enormity of what has actually happened and what has been lost. In time this depression is replaced with *acceptance,* as the person finally accepts what has happened.

These stages are universal whether we are mourning the deceased, the loss of a job, a child leaving home for good, or any other loss. Victims go through these same stages as they deal with the losses inflicted upon them by the violence.

There is no one medically necessary way to grieve so long as the loss is addressed. Some persons become very hostile toward a specific person who symbolizes the loss. For example, a victim may be furious with police personnel who should have been there to protect. Some victims become overactive and deal with the loss by keeping very busy. Some victims develop agitated depression or other forms of medical illnesses. Still others may become socially withdrawn or hide their feelings of sadness and anger.

Each victim must go through the various steps of the mourning process: accepting the reality of the loss, taking leave of it, and finding a new reason to go on. Unresolved grief takes its toll on human health and well-being. A recent study of all persons brought to a medical emergency room for reasons thought to be medical illness found that in as many as ten percent of those cases there was no imminent medical illness per se causing the patient's acute distress. These patients had unresolved grief over losses that they had not dealt with. For example, a colleague of mine recently treated a woman with chest pain in the emergency room. There was no medical basis for the pain, but it was the date of her son's death in the Vietnam War. This was a loss about which she had never spoken to anyone.

Victims need to share the pain with another person. Sharing the grief and its feelings allows victims to free themselves from the loss, to adjust to their newly limited environments, and to form new reasons to make life meaningful. Grief counseling freed the woman in the emergency room from her chest pain.

Personal Feelings and Traumatic Events. These disruptions in mastery, attachments, and meaning often result in an array of negative emotions or feeling states. Sometimes there is one predominant feeling, often the victim has several different feelings. Here are some of the more common emotions that victims often have, and that you may have noticed in yourself.

Anxiety is the first and very understandable reaction. It is normal to feel anxious when victims are confronted with a potentially life-threatening event. The body, as we have seen, is built to respond with anxiety. When the traumatic event has passed, victims may sometimes experience some of the other feelings listed below, but this is not always the case. When the traumatic event is the result of human maliciousness rather than a natural disaster, victims often remain anxious, as they fear the assailant may return again. This reaction may be especially true for persons who have been physically or sexually abused.

Anger is a second understandable reaction to traumatic events. Anger, like physical pain, is a signal system to alert victims to possible danger so that they pay attention and protect themselves as best they can. Since anger helps them to mobilize to cope, it often has an additional effect of reducing feelings of anxiety that they may be simultaneously experiencing.

Anger is helpful when it leads victims to solve the real problems that are before them. There is a continuing debate over whether it is better to express anger or to keep it in (Tavris, 1982). Which style persons choose seems to depend on their make-up and their life experiences. In either case it is more important to solve the problem that is causing the anger. For example, if a loved one is killed by a drunken driver or murdered in a drive-by shooting, the family's anger that leads to legal redress in the courts would be a constructive use of anger to address this painful issue.

Unfortunately, not all victims are able to directly confront their assailant. Not all rapists or street criminals or combat enemies are caught, and in natural disasters there is no human agent to blame. In these cases, victims may well be angry, and the anger about the event still needs to be expressed so that victims can resume more normal lives. It is important that victims discuss their anger about the specific violent event, and not focus the anger on other innocent people. Being continually angry with others will only isolate the victim from needed supports, as we have seen. Equally destructive is the stance of some victims who turn the anger on themselves (e.g., by arranging to fail often, by acts of self-mutilation, etc.), and who often end up in a state of self-hate. Both of these extremes are not adaptive solutions and need to be avoided.

A third common feeling is shame, which arises from the experience of being looked down upon by a group of other persons. Sadly, for many victims it is sometimes true that the larger community in its ignorance continues to blame the victim. This blaming of the victim is gradually lessening, but it is still common. Shame, however, can arise in victims from their own imposed, but faulty, self-assessments. Some victims feel shame when their minds or bodies have been invaded, such as in rape or torture, as well as in cases when families have past secrets of bankruptcies, suicides, childhood deaths, secret traumatic events, etc. (Fossum and Mason, 1986). The sense of self-imposed shame leads many victims to attempt to be perfect and to always be in control. Since no human can always be perfect or in control, shame-based victims continually experience failure, and the cycle of feeling ashamed begins again. In addition, the constant straining for perfection and control often leads to addictive behaviors as victims attempt to self-medicate their feelings of frustration and failure. Shame is common in many victims, especially in persons from homes with family alcoholism. Often the alcoholic family itself is governed by the rules of shame-based families that we shall discuss in chapter 7.

A fourth feeling state often found in victims is guilt. Victims frequently feel guilty about the violence. They feel responsible for the traumatic event, even though by definition such events are something which the victim has no control over. In the short term, victims may be using self-blame as a method of self-control. This is an acceptable place to start, but when guilt becomes a continuous state of mind, the victim is not well-served by self-blame.

Many victims that I have counseled have found it helpful to draw the distinction between moral guilt and psychological guilt. Moral guilt arises when someone consciously and willfully disregards the laws of God or of society. Moral guilt stems from intentional breaking of the rules. Psychological guilt, however, arises when a person is attempting to meet some superhuman set of standards that have been imposed by others. For example, battered women often come to believe that they deserve to be hit for not having kept the house spotlessly clean, or because one of the children has made a simple mistake. Children of alcoholic parents are often given superhuman goals by the drinking parent. I once had a patient in high school who obtained all As, and was captain of the football team. His alcoholic parent wanted to know why this teen was not also class president. Psychological guilt usually arises from superhuman standards that no person could ever attain.

The final common feeling is that of depression. While some forms of depression appear to be genetic (see chapter 4), many others often result from the loss of something important to the victim. The loss of personal integrity in battering, the loss of loved ones in homicide, the sudden death of a child may all lead to depressive feelings. Similarly victims, who also have lost their sense of self-esteem from any aspect of the traumatic event or its aftermath, may also experience depression. When victims feel depressed, nothing much in life is of interest and their meaningful purpose in life needs to be re-thought.

The Experience of Symptoms.   Let us return a second time to the example of Joan, the administrative assistant in the law office where the head of the firm and another lawyer had been arguing. In our first review of this matter, we were able to clarify our understanding of her "bad day" at work and at home by considering her past history of witnessing abuse between her parents.

In this second examination of her behavior on that day, let us consider how her early abuse history has led to psychological alterations in her sense of mastery, attachment, and meaning in her adult life.

We learned that Joan prided herself on methodical self-control, and that she took the job because it seemed manageable. After the argument had occurred, we are told, she was unable to organize the rest of her day. These themes suggest a victim using excessive mastery by attempting to control everything. Such victims hope to ward off any further sudden, unexpected, potentially life-threatening events. Joan's disrupted day is a good example of how such attempts fail.

The vignette also provides us with some sense of Joan's capacity for caring attachments. She is described as shy and as a loner. She took the law office position because it would require minimal interaction with others. She is and has been married for some time, so she has the necessary basic interpersonal skills, but she was unable to utilize the potential support her husband tried to offer her. It was easier to flee. Each of these examples reveals the subtle ways in which her history of traumatic abuse resulted in disruptions in the caring attachments that are so important to each of us.

Finally, we learn that Joan has taken the law position because she really doesn't know what she wants to do with her life. While this aimlessness may be a function of a late identity crisis, it is not unreasonable to assume this unresolved identity crisis and her general view of life are also impaired by a long-standing inability to grieve and make meaningful sense of what happened to her as a child.

Disruptions in reasonable mastery, caring attachments, and meaningful purpose that arise from traumatic events are not over when these events have passed. Their capacity to impair our lives remains with us until these matters are directly addressed.

*   *   *

It is remarkable how the medical research on the skills of stress-resistant persons and the medical research on the disruptions experienced by victims in traumatic events are so symmetrical and diametrically opposite. In spite of these painful traumatic alterations and intense negative feelings, there is a message of hope for victims. The skills of stress-resistant persons can be learned by any of us, and that includes victims.

Victims have learned to reinstitute mastery by first raising their children, returning to work, and even by balancing their checkbooks. Armed with these basic mastery skills, they then address the PTSD itself. Victims have learned to develop caring attachments by learning to trust other special persons, by relying on the clergy and counselors, or by joining self-help groups with other victims. Victims have begun to find new meaning in life by helping others. Natural disasters as well as other forms of traumatic events have produced remarkable examples of victims helping victims.

Again, each of you can learn to use these same kills for your own recovery, and we shall explore how to do this later on.

As you might expect, traumatic events impact not only on our psychological functioning, but on our body chemistry as well. Some of these biological changes occur instantly at the time of the traumatic event; others appear to emerge as long-lasting changes to the nervous system if the traumatic episode goes untreated. We turn to these biological changes next as we continue to pursue our understanding of posttraumatic stress disorder.

3

Why Are My Nerves So Frayed?
The Biology of
Posttraumatic Stress Disorder

Nothing fixes a thing so in memory
As the wish to forget it.
— Montaigne

Rage, rage against the dying of the light.
— Dylan Thomas

Teacher of the Year. No small feat in an era of shrinking budgets and expanding class size. Third-graders were not easily taught in the best of circumstances, but thirty-something, attractive Miss Sarah Radcliff was considered an outstanding, preeminent teacher and citizen of this small rural southern Vermont community. Parents were impressed with her zeal for learning that took her to Albany, New York, several nights each week for continuing education courses. She felt both pleased and fraudulent. But the great ache which kept her constant prisoner was not assuaged.

It had begun when she was four. Her father had come to her bedroom when her mother was at church. He made her undress, and during the first few times he lay quietly beside her. But it soon progressed. She felt guilty because initially it felt comforting to lie beside human warmth, however frightened one might be. But repeated oral, anal, and forced vaginal entry quickly overrode any sense of earlier comfort. He threatened her with death if she ever

44

spoke to anyone of this. Toward the end he beat her with a hair brush at the start of each encounter. This occurred every Sunday morning of every week for eleven years.

Now years later, as Sarah rolled back into the covers, she absorbed the glow of climax, a warmth enhanced by scotch. She found sex comforting, soothing somehow; yet it was short-lived and left her feeling empty. Three lengthy affairs with married men, more one-night stands than she could count.

She hated it. She hated the desire to have sex with every man she saw. She hated cruising the bars in Albany. She hated drinking too much. She hated feeling like a piece of picked-over meat. And yet she could not stop. Something drove her to this again and again and again. More than once she thought of ending it all.

Teacher of the Year. Every Sunday morning of each week for eleven years. Could loneliness kill? Would her great heartache be her life-long partner until death parted them? There was only silence.

*   *   *

Sarah is a victim of repeated abuse—in this case repeated incest by her own father, who has betrayed her trust. We can see in her behavior the legacy of the psychological disruptions that we spoke of in the last chapter. Her frantic need to remain busy to feel in control, her inability to form close relationships to others, and powerlessness to find some meaning in life that would lessen her despair. In her quiet, often heroic way, Sarah, like many victims, is attempting to carry on with life in the midst of great pain.

Sarah's current life, however, has additional lessons to teach us about the untreated consequences of PTSD. Her inability to relax, her continuous sexual desire and liaisons, her abuse of alcohol, her depression, and her intermittent thoughts of suicide may well be related to changes in her body chemistry as a result of the abusive episodes in her childhood that went untreated.

Behavioral science is learning that the intense biochemical changes that occur in the victim at the time of the traumatic event may lead to permanent alterations in the victim's nervous system. These alterations may be linked to any of the PTSD symptoms that may appear later on in the victim's life, as well as to the repetition compulsion and addictive behavior. Without treatment and recovery, past untreated trauma may become a true chronic medical illness.

Victims often find these symptoms, repetitive compulsions, and addictive behaviors to be a puzzlement. The victim's experience of these symptoms and behaviors is often one of sudden onset at unpredictable times for no known reason. It is as if the body speaks its own PTSD foreign language. This view, however, is not accurate. The emergence of symptoms and/or compulsive and addictive behaviors at various times is predictable, is understandable, and can be controlled once the victim understands the links between these problematic events and the current biological consequences of his or her past and untreated traumatic event(s).

In this chapter we shall explore these biological changes, which occur in all trauma victims. We will review the emergency response systems that are there to protect us when violence occurs, and then outline the fairly permanent changes that may occur in the PTSD aftermath in the absence of treatment. This chapter will help you understand the foreign language of PTSD, and will help you understand your feelings, symptoms, compulsions, and addictions in an entirely different light. They are rooted in biochemical changes that are understandable and that can be controlled once you understand the patterns. This chapter is especially important for those of you with addictive behavior as it will show you the possible biological roots of your addiction that may have developed shortly after your abuse.

We shall close the chapter by asking an important question: Who develops PTSD? Not every victim of a traumatic act of violence goes on to develop PTSD; in fact, many appear to emerge from these dreadful events relatively unscathed. The initial research on this question is of interest to all of us, and we shall examine these latest findings to see what they can tell us about these two different outcomes.

Let us begin with a brief review of how the human nervous system responds to the severe life stress of psychological trauma and PTSD.

How the Nervous System Works

The Nerve Cell

The human body is composed of billions of cells that form all of the parts of our body. This includes the nerve cells which allow us to control our bodies and to respond to the world around us.

FIGURE 1

Synaptic Transmission Between Nerve Cells:

Synaptic Gap
(with Neurotransmitters)

↓

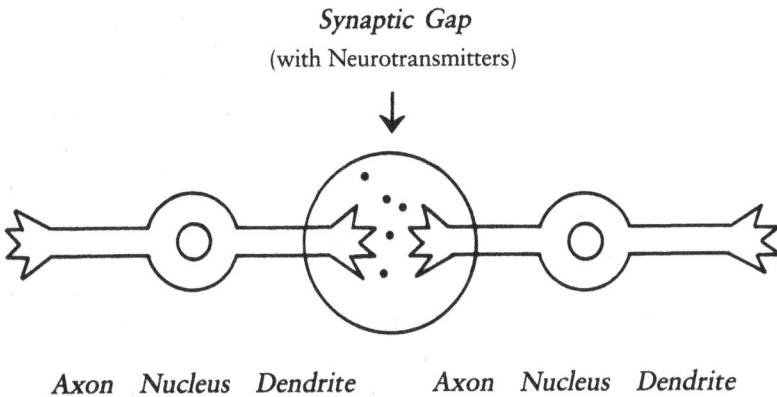

Axon   Nucleus   Dendrite   Axon   Nucleus   Dendrite

Each nerve cell has an axon, a cell nucleus, and a dendrite. If you begin to tap your thumb on a desktop, the sense receptor under your skin in the thumb transmits the message along the axon part of the nerve cell. The axon carries the message "desktop" to the cell nucleus. The nucleus then transmits the same message along the dendrite to the next nerve cell's axon. These bits of information or messages are transmitted via bio-electrical impulses along the pieces of each nerve cell. This process is outlined in figure 1.

The human nervous system is not composed of long uninterrupted lines of nerve cells. Rather the nervous system is a series of pieces of nerve fiber, composed of many cells operating in the manner we have just outlined. The information from the sense receptor is passed from one piece of nerve fiber to the next until the message reaches the brain, and registers as "desktop."

If the nerve message pathway is a series of nerve fibers, and not one continuous piece, the question arises as to how the message gets to the brain? How does the message get from the dendrite of one nerve fiber to the axon of the next nerve fiber, and so forth on up to the brain so that we are aware of what is happening?

The answer lies in a small, microscopic balloon-like sac called the synaptic gap. As we can see in figure 1, this synaptic gap encircles the dendrite of one fiber and the axon of the next fiber. These synaptic gaps or balloon-like vesicles hold several chemicals. These chemicals are called neurotransmitters, and when they are in proper balance, the chemicals act like a drum head, and vibrate or transmit the message from one nerve fiber to the next. These messages are transmitted clearly, and the person functions efficiently. This transfer of information within the body is analagous to a large telephone network. As long as all of the circuits and transmission lines are working correctly, calls between persons continue uninterrupted.

Psychological trauma, however, alters some of these neurotransmitters in the victim's body and in the victim's mind. Information does not flow smoothly, and the victim does not function optimally in the traumatic crisis.

Here is a list of the five neurotransmitters that seem to be clearly altered in the face of traumatic events:

1. *Epinephrine.* Epinephrine, a by-product of the adrenal gland, mobilizes the body to cope with the stress of the traumatic event. It regulates and mobilizes heart rate, breathing, muscles, blood sugar for energy, and the like so that victims are prepared for the crisis at hand.
2. *Cortisol.* Cortisol is also released by the adrenal gland. In times of danger or uncertainty, it provides a source of energy by releasing blood sugar into the bloodstream and repairs body tissue in cases of injury.
3. *Norepinephrine.* This neurotransmitter is also a by-product of adrenaline and is transmitted via the bloodstream to the brain. Norepinephrine acts as a general facilitator in the brain to enhance alertness and efficient problem-solving.
4. *Serotonin.* Serotonin is a transmitter produced in the brain itself. When it is present and in adequate supply, the person is calm, relaxed, and generally content. When serotonin is depleted, the person becomes irritable, angry, unhappy, and ultimately depressed.

5. *Endorphins.* Endorphins are also found in the brain. When they are actively circulating (e.g., during aerobic exercises), the person feels calm and relaxed, and often has a sense of well-being. A decrease in endorphin circulation may leave an individual irritable, angry, and unhappy.

These neurotransmitters are centrally involved in the body's and mind's emergency response systems, and in the subsequent development of the various symptoms of PTSD and addictive behavior. They are the keys to understanding the foreign language of PTSD.

The Cortex and the Limbic System

To understand psychological trauma, it is important to be aware of the cortex and the limbic system as well as nerve cells.

The cortex is our conscious, thinking brain at its highest level of functioning. Information comes from our senses (hearing, taste, touch, smell, sight) to the cortex for processing. Memory is scanned for past similar events. Decisions and judgments are made and implemented as the cortex responds to whatever issue has presented itself and must be coped with.

The limbic system lies in the center of the brain under the cortex, and it is about the size of a thimble. This small part of the brain plays a very large part in our experience of the world. It adds the dimension of feeling to our life experiences. All messages coming from the sense receptors in the body pass by the limbic system on their way to the brain for processing and response. It is the limbic system that guides our emotions as well as the instincts for survival that include feeding, fighting, fleeing, and sexual behavior. Fear, anger, peace, joy are all in part determined by the nerve cells and neurotransmitters in the limbic system. Both the cortex and the limbic system play a crucial role in our experience of the terror of trauma and its PTSD aftermath.

The Emergency Response System: Mobilizing the Body

Trauma mobilizes the victim to hypervigilance within seconds so as to ensure survival. Here are the basic components of the process.

The victim's brain begins with a two-step cognitive evaluation of the problem at hand (Lazarus and Folkman, 1984). The brain makes two appraisals instantly: "Is this a life-threatening situation or not?" and "If it is, what can I do to cope to survive?"

If the brain judges the situation to be life-threatening, the body instantly responds with the second component, which is referred to as the "fight or flight" response. Physician Walter Cannon (1963) studied this body response for many years, and has clarified it for all of us. The body's emergency system is mobilized to respond when the neurotransmitters epinephrine and cortisol are released. This leads to a strengthened heart rate, better respiration, dilation of pupils for better vision, the release of sugar into the blood for greater energy and less fatigue, and the enhancement of the body's capacity to coagulate blood and repair any damaged tissue. The individual undergoing these changes may feel tension, trembling, spasms, or dizziness as the body prepares to cope. And all of this happens without our ever having to think about it.

Physician Hans Selye (1956) noticed that this physiological response had its own internal three-stage process which he called the General Adaptation Syndrome. The first stage is the alarm or readiness stage that we have just outlined. In the second stage, the body works to solve or resist the problem before entering the third stage, or recovery process, where the body returns to its normal resting state and restores its energy for the next crisis. Untreated psychological trauma disrupts this normal adaptive pattern of readiness, resistance, and recovery because the victim's hypervigilance and distress prevent the body from entering the recovery process, which would allow the person to return to a more normal pre-crisis resting state.

Notwithstanding, the physiological activation of the body is a truly remarkable system of protection, and we are always mobilized in this fashion when we are faced with traumatic events.

The Emergency Response System: Mobilizing the Brain

When the brain makes its appraisal of danger, there is a second and simultaneous emergency mobilization response in the victim's mind. This mobilization of the mind or brain involves the cortex, the limbic system, and the neurotransmitters norepinephrine, the endorphins, and serotonin. The recent studies of the role of neurotransmitters in the victim's brain biochemistry are among the most exciting advances in modern medicine, but these changes are very complicated, and scientists are still hard at work trying to clarify exactly what happens. The following section presents our current understanding of the brain's emergency response system in the face of traumatic events. Readers with a fundamental interest in such biochemistry may wish to read further in this chapter's Select Readings. The presentation outlined here draws largely on the work of psychologist John Wilson (1989), and psychiatrist Bessel van der Kolk (1987; 1996).

The Brain's Basic Response

When the person is confronted with the traumatic event, the victim's conscious, waking brain responds with increases of the neurotransmitters norepinephrine and endorphins in the cortex and the limbic system.

Norepinephrine alerts the whole brain to the crisis at hand, and the cortex is mobilized to think out how best to respond to ensure survival. About thirty seconds after the norepinephrine is released and in place, the brain enhances the circulation of endorphins, those neurotransmitters that seem to make us calm and relaxed.

At first, this release of chemicals that relax us was a puzzle to researchers. It would seem that none of us would want to be relaxed and casual during a life-threatening crisis. Continued research, however, has begun to reveal the wisdom of Nature during these crises. The release of endorphins appears to serve at least two functions. First, these neurotransmitters facilitate thinking and remembering clearly, and help the victim remain relatively calm. Secondly, endorphins appear to act as analgesics or pain killers for about an hour and a half. Some examples: victims in a car accident may walk away from the flaming, crushed vehicle, and begin to experience pain from sustained injuries only when they are safely removed from harm's way. Similarly, soldiers in combat may not feel the pain of shrapnel until they are safely back at camp. One of my colleagues treated the mother of a child who was hit by a bus. The mother accompanied the child to the emergency room and calmly told the doctors what had happened and what she thought needed to be done. When the doctors emerged from surgery two hours later, the mother was visibly trembling, crying uncontrollably, and was ashen white. Such temporary dampening of pain and enhanced clarity of thought appear to be Nature's way of protecting us so that we can concentrate on saving our lives first and foremost.

Physical Symptoms. We noted earlier that any of the three groups of trauma symptoms can begin during the crisis, or in either the acute or chronic phases of PTSD. The neurotransmitters in the cortex and limbic system are involved in the development of these symptoms.

The increased presence of norepinephrine which alerts the brain also appears to be involved in the onset of the physical symptoms. Hypervigilance, an exaggerated startle response, fright, panic, anxiety, insomnia, and mood irritability all appear in part to be a function of increased amounts of norepinephrine.

Intrusive Symptoms. Similarly, the presence of both norepinephrine and endorphins appears to be involved in the development of the intrusive symptoms—intrusive memories, flashbacks, and nightmares.

Norepinephrine and endorphins together appear to produce the best brain chemistry for learning and remembering. With these two neurotransmitters present, the traumatic experience and the way we respond to it are etched or "burned" into our memory. It is Nature's way of saying: "This is a very dangerous life-threatening situation, and don't ever forget it in case it should happen again." This chemistry of indelible learning and remembering probably contributes to the difficulty victims have in resolving the various types of intrusive symptoms.

Just as excessive norepinephrine can produce troublesome physical symptoms, excess levels of endorphins can result in faulty learning, poor memory, or possibly amnesiac states. While amnesia in victims may also be caused by head trauma, post-concussion syndrome, or general mental deterioration, excess levels of endorphins may be involved in some situations in which the victim can remember nothing of the traumatic episode. Some researchers believe endorphins may be involved in the flashback phenomenon related to dissociation. They theorize that the endorphins produce some part of the analgesic or even amnesiac effect that allows the brain to put the consciousness of the event on hold.

Avoidant Symptoms. Finally, a particularly severe traumatic episode, or repeated episodes of abuse decrease the presence of norepinephrine and the endorphins in the brain of the victim. In addition, severe traumatic stress also causes the neurotransmitter serotonin to be depleted. The depletion of these three neurotransmitters results in the onset of the avoidant symptoms. Victims avoid activities, avoid other people, and generally withdraw into isolation. They have little motivation for anything, generally find little pleasure in life, and often become depressed.

Learned helplessness, which is found in victims of repeated abuse, very frequently leaves helpless persons depressed. There is speculation that continuous increments in arousal coupled with decreases of norepinephrine, the endorphins, and serotonin may lead to a biochemical state of helpless depression. The depression associated with helplessness, however, is different from depression in nontraumatized persons. In coping with general life stress, epinephrine, cortisol, norepinephrine, and endorphins are released. When the event passes, these neurotransmitters, along with serotonin, decrease, and the nontraumatized person may become depressed. In victims of trauma who become depressed, reduced cortisol is present. No one knows why. Repeated abuse, such as incest, usually occurs in childhood, and it is thought that the reduced cortisol must serve some adaptive purpose in ways that are not yet known. While this process is not yet fully understood, there are effective treatments for depressed victims with PTSD.

Kindling. Kindling refers to a phenomenon in which small amounts of norepinephrine may produce an emergency response as intense as the original traumatic event. These post-trauma arousal states seem to be related to seemingly permanent changes in the nerve fibers of the victim's limbic system, which is involved in producing the feelings victims experience. Repeated exposure to

stressful traumatic events leads to continuous exposure to norepinephrine for vigilant alertness. The continuous presence of
norepinephrine appears to change the nerves in the limbic system
in such a way that little events that would normally be shrugged
off as minor stressful situations suddenly produce a full-blown
case of "frayed nerves." Your child's skate on the floor, the other
driver's carelessness, a sudden sound in the back of the room—all
of these seemingly inconsequential events can lead to full hypervigilance, startle response, anger, and irritability in some trauma
victims. The victim essentially develops a case of chronic overarousal, and can never "relax."

For the victim with kindling and its state of easily induced
overarousal, *both* the negative and positive events of life may produce the small increments of norepinephrine that bring about
the unpleasant arousal and vigilance. Unpleasant events like arguments, traffic jams, waiting in long lines can produce the unwanted emotional response; but so also can pleasant events like
greeting others, parties, sporting events, sex, the content of movies, and so forth. It is not uncommon for the victim to attempt to
gain control of the unpleasantness by blunting all forms of emotions. The most commonly employed strategy is to dampen these
feeling states by avoiding people and places as much as possible.
This can include solutions like driving in non-peak hours, going
to the supermarket later in the evening when the crowd is gone,
or going on vacation in the Fall to a secluded beach when most
people have returned to work. While this approach works for
some victims, their lives remain constricted, and they do not avail
themselves of the better treatments for this problem.

The Experience of Symptoms.   At this point in our discussion
of the biological basis for some victims' behavior, let us return for
a third time to the example of Joan, who had a past history of
witnessing physical and verbal abuse between her parents, and
who became markedly upset when her boss and another lawyer
had an intense verbal argument.

You will remember that Joan was distressed all day, had a headache by mid-afternoon, was forgetful at home during the dinner
hour, and had an argument with her husband before she went to
her room to be by herself for the evening. Let us review her situ-

ation here again so that we may now understand some of the possible biochemistry in her body and her brain when the event occurred and as her day wore on.

The argument between her boss and the other lawyer appears to have been a symbolic reminder for Joan of the violence she witnessed between her parents when she was a child. The event in the law office may have re-created body and brain chemistry similar in Joan to the body and brain chemistry she had as a child when the family violence occurred. Epinephrine, cortisol, norepinephrine, and endorphins were released.

The remainder of her day was marked by "distress." Some of this distress was related to her body's emergency mobilization, with its release of epinephrine and cortisol so that she became restless and apprehensive, and had a headache by the afternoon. Some of this was due to the norepinephrine and endorphin activation in her brain. She had trouble concentrating at work and was "forgetful" at home in the evening. Joan was essentially having physical and intrusive symptoms. Her husband's offer of assistance overwhelmed her. The day's hours of arousal plus his offer and the fact that she would have to interact with him had led to decreases in norepinephrine, the endorphins, and serotonin. At this point Joan was experiencing avoidant symptoms, and she went off by herself to blunt the feelings of arousal, and to allow her body and brain to recover.

Since Joan was not in a treatment program at the time, both her overt behavior and her internal body and brain chemistry were automatic responses to events. She was unaware of what she was doing and how it may have been related to her past history of family violence. She was aware, however, that she felt miserable.

Victim Distress and Addictive Behaviors

We have seen how severe traumatic stress may produce kindling, so that small amounts of norepinephrine can initiate the cycle of PTSD symptoms. This process is further complicated by the depletion of the endorphins that lead to the avoidant symptoms in victims. When the traumatic crisis has passed, the endorphins decrease along with norepinephrine and serotonin. Victims may enter endorphin withdrawal somewhat similar to opiate withdrawal, and may experience withdrawal symptoms such as restlessness, agitation, some minor trembling in the body, and other flu-like symptoms. To make matters worse, many victims become frightened when this happens. This results in an increase in small amounts of norepinephrine, kindling occurs, and victims begin to experience the onset of physical and intrusive symptoms. It becomes a vicious cycle.

How do most untreated victims cope with this endless cycle? There appear to be some common approaches.

First, the victim can allow the emotional arousal to increase to the point where the victim's current brain chemistry is fully similar to the norepinephrine level that was present at the time the original traumatic event occurred. It is thought that this re-creation of the original state enhances memory retrieval and resolution of the problem. Theoretically, at least, the victim could then reexperience the intrusive memories and begin to work at assimilating the painful event to try to master it. Such an approach, however, is fraught with difficulty. If the victim reexperiences the whole intrusive memory all at once, he or she runs a high risk of being psychologically overwhelmed and biochemically retraumatized. A victim's recall needs to occur in small and manageable steps, and a strategy of increased arousal will not necessarily permit a graduated approach to the traumatic material.

A second strategy is for the victim to self-medicate the unpleasant withdrawal symptoms by engaging in some form of addictive behavior that will enhance endorphin presence in the brain. Sexual addictions, fist fights, binge eating, self-mutilation, cocaine and crack, reckless driving, and the like appear to produce a short, intense increase of endorphin release. These effects will last about one-and-one-half hours, and will temporarily "treat" the withdrawal symptoms. The problem with this strategy is that the victim will seek yet another endorphin release as the first one wears off. Over time, the addictive behavior becomes worse and the trauma remains untreated.

A third strategy that a victim might employ is an attempt to blur the pain of the withdrawal state with alcohol or drugs. This too is a form of self-medication with the goal of sedating the distress. The use of alcohol and drugs like the minor tranquilizers for anxiety, or barbituates for sleep, is another ineffective solution

FIGURE 2

The Biochemistry of Traumatic Events:

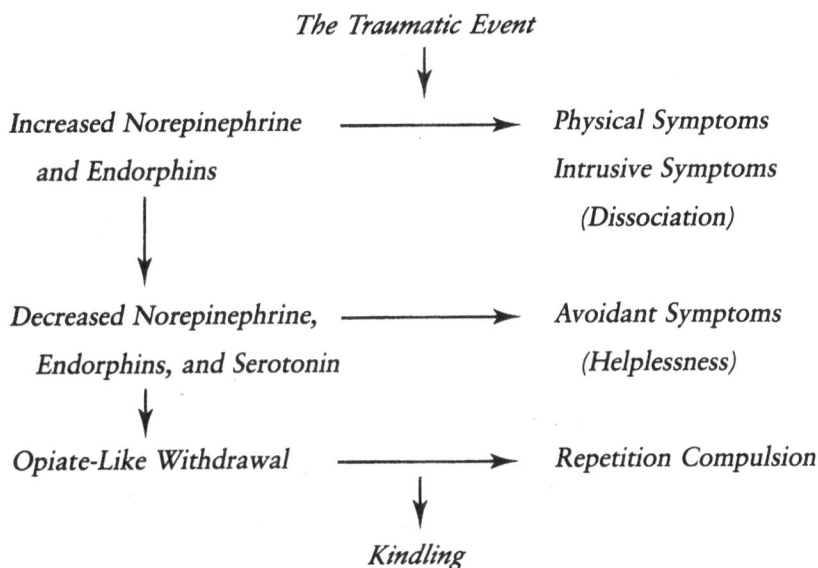

The Traumatic Event

↓

Increased Norepinephrine ⟶ *Physical Symptoms*
and Endorphins   *Intrusive Symptoms*
                                              (Dissociation)

↓

Decreased Norepinephrine, ⟶ *Avoidant Symptoms*
Endorphins, and Serotonin   *(Helplessness)*

↓

Opiate-Like Withdrawal ⟶ *Repetition Compulsion*

↓

Kindling

---

because the victim is again at increased risk for ending up with an addiction.

A fourth and better strategy is to alter the biochemistry of traumatic events in ways that are adaptive and do not result in addictive behavior. As we shall see in the treatment and recovery section, some prescribed medicines, relaxation exercises, and aerobic exercises are all helpful alternatives.

The Biochemistry of Traumatic Events Revisited

Figure 2 presents a brief overview of the biochemistry in the brain that we have reviewed in this chapter. The traumatic event occurs, and the neurotransmitters epinephrine and cortisol mobilize the body's emergency response system. At the same time, norepinephrine and the endorphins mobilize the brain's response, and lead to the physical and intrusive symptoms, (including dissociated experiences). As the norepinephrine, endorphins, and serotonin decrease, the avoidant symptoms emerge and may well include depression and helplessness. Prolonged stress or repeated abuse may create endorphin opiate-like withdrawal in victims, and may result in repetition compulsions and addictive behavior. Prolonged stress or chronic abuse may also ultimately lead to the permanent changes in the limbic system that are referred to as kindling.

The following example is a helpful way to remember the various types of biochemical changes that may occur.

*   *   *

"Freeze, Olsen." Olsen froze, and was simultaneously forced to the floor by one of the corrections officers on the scene. This was the third time in a month that Olsen had struck another inmate, and he had served only five months of his four-year term. "Are you paranoid? Do you always hit people who touch your shoulder?" Olsen shrugged, and quietly returned to his cell and the concrete solitude that had become his life.

Fighting was no stranger to Olsen. It was his closest companion, and the reason he was incarcerated. Eight months earlier, he had been minding his own business at the tavern, drinking his beer, and watching the football game until someone came up behind him to challenge him to a game of darts. In one impulsive split second he had flattened the other patron. One thing led to another. Glassware was broken. The police were called. As the arresting officer, with his gun drawn, reached behind Olsen to handcuff him, Olsen again exploded in rage. His sentence: four years in jail for assault and battery of a police officer.

Jim Olsen had always had a chip on his shoulder. His fellow workers at the body shop knew it; his wife and children were equally aware. He had been a street-smart kid who stole cars, and could drive with the best of them by the time he was twelve. Stealing money was his specialty. Gas stations, convenience stores, little old ladies, his own mother. Society owed him—especially for Vernon Sparhawk, the drunk.

Vernon had lived by himself, three doors down, in the housing project. Vernon was the neighborhood drunk, but Mrs. Olsen, a health aide, was determined to raise her kids right. She insisted that her children treat everyone with respect, and everyone included Vernon Sparhawk.

When James was eight years old, Vernon had asked Jim to bring him back a loaf of bread from the store. Jim did as he was told. When he returned, Vernon grabbed him at knife-point, tied him to the headboard of the bed, whipped him with a leather belt until James was bleeding, and then sexually penetrated him from the rear. This was to happen four more times within the month. Each time James was threatened with death if he told.

Jim Olsen, the child and adult, had never told. Frightened and humiliated at the time, he remained filled with shame and hurt, but it became a point of honor never to tell. Big boys don't cry. His anger, however, had now become the infamous chip on his shoulder. To this day, when anyone came up behind him, he had flashbacks of Vernon Sparhawk, the alcoholic in apartment 7B.

*   *   *

Jim Olsen's behavior demonstrates the various aspects of the biochemistry of the traumatic response. We learned that Olsen was constantly vigilant and easily startled (physical symptoms), and that his two emergency mobilization response systems went into full alert when the patron approached him for a game of darts. Memories of his past childhood abuse at the hands of Vernon (intrusive memories) were probably stirred when the patron approached him from behind, and again, when the police handcuffed him. The chip on his shoulder and his poor relationships with others reflect his social isolation (avoidant symptoms), and his repeated antisocial acts and pattern of abusive drinking may reflect, in part, the phenomenon of kindling and some of the long-term consequences of his untreated PTSD.

Who Develops PTSD?

Before we close this chapter, we need to consider one further and important question. Are some of us more likely to develop PTSD than others? Clearly, not everyone who experiences a traumatic event then goes on to develop PTSD. How are we to understand this? Is it a function of the biology or personal make-up of some persons that results in a better ability to respond to the event, or is it something in the quality of their personal life experiences?

This question is known as the nature/nurture argument, and it has a long tradition of study in behavioral science. When a person responds to a life event, what is more important in determining the person's response? If we believe biology is the most important factor, then we support the nature side of the argument. If we believe the resources in the environment are most important, then we believe in the nurture side of the argument. For example, if someone is really very bright and creative, is the person's high intelligence a function of the brain the person was born with (nature) or the excellent schooling he or she has had (nurture)? Similarly, in our own case here, if a person develops PTSD, is it because the victim was born more vulnerable (nature) or was the stressful event itself too overwhelming (nurture)?

Medicine has usually resolved the nature/nurture question by relying on twins studies. In twins studies, investigators use identical monozygotic twins so that they know the biology is the same for each twin. In addition, they study only monozygotic twins reared apart since birth. They try to select those cases in which both twins seem to have been reared in home environments that were highly similar. The basic research strategy is as follows. Since the biology of the twins is the same and their home environments are as similar as possible, if the twins who have been reared apart end up having the same medical problems as each other and as their biological parents, then the particular illness under study is probably genetic. Nature would be the important factor. If the reverse is true, and the twins have different health problems than each other or their biological parents, then the home environment of the adopting parents must exert some influence on the development of the particular illness under study. Nurture would assume the more prominent role.

Exacting twins studies have been done on two major illnesses—schizophrenia and depression—and in each illness *both* factors have been shown to be important. The inherited biology of the person interacts with stressful life events, and the interaction of the two contributes to the onset of a particular disorder.

Some of the PTSD research to date has found some evidence to suggest a biological vulnerability in some people who become victims (nature). Other studies have shown that severe family stress in childhood (nurture) may leave individuals likely to develop PTSD in the face of subsequent traumatic stress. Family problems such as unemployment, alcoholism, divorce, arrest, and parental fighting have all been cited as possible contributing factors. Prior trauma, hyperarousal, and lack of social support also seem important (Litz, Gray, and Adler, 2002).

In truth, no one knows, as yet, why some people develop PTSD and others do not. In time we may learn that it is a function of a person's biological vulnerability in interaction with the stressful events in that person's life. This has been the case with some other medical illnesses, as we have seen, and it may prove to be true of PTSD. Future research will be needed to answer this complex question, and while we wait for these findings, it is important for all victims to remember that there are successful treatments for recovery that are available now.

* * *

This chapter has focused on the biochemistry of the emergency response systems in the body and brain when confronting traumatic events, so that we might better understand the foreign language of PTSD and how its symptoms, the repetition compulsions, and some patterns of addictive behavior may be systematically understood and even predicted. Such knowledge increases a sense of true mastery in victims, decreases their shame, and leads to specific strategies that victims can utilize for recovery.

One very common method for victims is to use a self-help program to attain sobriety or remain straight, if this is necessary; and then to use relaxation or aerobic exercises to contain and diminish the symptom cycle associated with kindling. With this initial treatment for the biochemistry of posttraumatic stress disorder, the victim can then address the remaining steps for recovery from the aftermath of PTSD. This has been a successful approach for many, and one that you can include in your own recovery program, if need be.

We have one last general area of study to complete our general understanding of the aftermath of PTSD, regardless of the specific type of violence encountered. It is a review of the major health problems that may arise when traumatic events and their PTSD aftermath are ignored.

4

It Hurts. The Several Faces of Untreated Traumatic Events

Sick men have no holiday.
— Camus

Even the desert blooms.
— Anonymous

She had awakened that morning in a state of cold rage. It had come upon her suddenly and without warning as it always did. Ethel never knew what started it, but she did what she always did to cope with it. She would speak to no one. She would go directly to work and distract herself by filing insurance claim reports. If she were lucky, the distress would pass by noon. She lit a cigarette, and paced restlessly as she waited for the bus. She was a walking ball of fire.

Anger ate at her concentration, and her attempts to lose herself in her work did not help. At lunch she had three vodkas and water to calm her nerves. Usually she started the day with some cocaine because it made her feel less depressed, but on the days of rage like today she needed alcohol to calm herself down.

By three o'clock she'd lost it, and found herself in a verbally abusive, no-holds-barred screaming match with her supervisor. Arguments like this one had happened before and appropriate penalties had followed, but today she was fired. There was only one way to stop the hated range now. She walked the seven miles to home. Screaming at drivers who slammed on their brakes as she crossed intersections against the light, and bumping carelessly into pedestrians on the sidewalk, she went home to do what she had to do. To free herself of the demon rage.

When the deed was done, calmness returned. She picked up the phone, dialed 911, lit up a cigarette, and waited.

Deborah Wilson, the triage nurse in the emergency room, had seen this frequent visitor before. She examined Ethel's wrists and forearms. Eight delicate self-cuts on each arm and wrist. Probably a razor blade again. Blood was seeping freely, but no major artery or vein appeared severed. Deborah was worried. Delicate self-cutters like Ethel had about twenty-two chances at this sort of thing. After twenty-two separate episodes of cutting, the scar tissue could be so extensive that any subsequent cut could prove fatal.

Ethel knew the routine, and she felt no pain as the physician sutured each arm. Next would come the psychiatrist with questions about harm to self or others. She would deny such thoughts, the hospital would release her, and this miserable cycle would begin again. It would be four more years before she would learn that the sudden cold rage that she could not explain was linked to her being sexually abused by her father when she was four years old.

*   *   *

Baseball legend Yogi Berra has a saying that it isn't over till it's over. This is a helpful expression for untreated victims and their loved ones to remember. Even though the traumatic event has passed, and the victim's life has returned to some degree of apparent normality, the probability of subsequent physical and psychological distress remains high.

One recent study of over two thousand female victims of various types of abuse found that *every single victim* saw a physician for some trauma-related health problem within two years of the violent episode. Still another study found that combat veterans who had PTSD associated with their active combat experiences had more chronic illnesses later in life than their noncombatant peers. Such studies show us that untreated violence may have long-standing health consequences.

We have seen how untreated psychological trauma may evolve into posttraumatic stress disorder. In this chapter, we shall learn how PTSD, if it is left untreated, may lead to even more serious medical conditions. Some of these conditions may severely limit the victim's capacity to function from day to day, and result in constant misery.

Ethel is a good case in point. She was a victim of violence that was left untreated, and her impairments have now resulted in a medical illness called Borderline Personality Disorder. Ethel's rage, social isolation, and self-mutilation are common characteristics in persons with Borderline Personality Disorder. Ethel also shares something else in common with them: a past history of sexual abuse. Over thirty percent of persons with Borderline Personality Disorder have a past history of some form of sexual abuse (Herman, Perry, van der Kolk, 1985). Like Ethel, they may fully remember the past abuse; and like Ethel, they have not drawn the links between their past abuse histories and their current functioning.

In this chapter we shall review the various medical conditions in which there are uncommonly high representations of untreated trauma victims. We shall begin the chapter with a discussion of the risk factors potentially present in any episode of violence that can make the medical outcome worse. We will then review those medical conditions where a past history of untreated abuse is often present. They are divided into four groupings: the anxiety disorders, the depressive states, the various addictions, and the personality disorders.

No doubt some of you may have heard of these illnesses and medical complications in other contexts, but here I am writing for you as victims. It is important for you to understand how these medical conditions that we shall list may be linked to untreated trauma and posttraumatic stress disorder. This review is not meant to frighten you nor to encourage you to make a self-diagnosis of your problem. Your family and friends, your physician, health care counselors, and the clergy can all help you to make sense of your specific circumstances. What is important here is that, if you have any of these problems in your own life, you want to begin to think about the possibility of their being linked to your past painful episodes of abuse, and to think about your needs for treatment and recovery.

An important lesson for all of us to take from this chapter is that time does not always heal the wounds of natural disasters and acts of human maliciousness. To the contrary, the pain persists, and often becomes worse. It isn't over till it's over, and it isn't over till it's treated.

The Risk Factors in Psychological Trauma

Traumatic evil acts do not occur in a vacuum. Such events take place in a context (Koss and Harvey, 1991; Wilson, 1989). The context includes the characteristics of three general factors: the person who is the victim, the specific act of violence, and the situation where the event takes place. It is this person x event x environment interaction that accounts for the wide variety of reactions to and recovery from traumatic events. For example, a female college student who is familiar with the feminist movement and knows of community resources for help will most likely react to and recover from a rape very differently than a socially isolated female whose community lacks such resources and is more prone to victim-blaming. Depending on the nature of the characteristics we are about to discuss, the victim may be at greater or lesser risk for recovery, and the presence or absence of all of the following characteristics needs to be assessed in each case of traumatic abuse. Psychologists Mary Koss and Mary Harvey (1991) have presented a useful listing of these factors.

Person Factors

Biological Predisposition.   Some individuals may be born with a predisposition to become easily aroused and to remain in chronic states of such arousal. It seems reasonable to assume that such individuals will experientially respond to evil acts of violence more intensely than victims without such innate chronic arousal. The course of recovery for the victim with chronic arousal becomes equally problematic. Many events in the environment will likely elicit the chronic arousal state and controlling the symptoms of PTSD will become more difficult and time-consuming.

Age and Stage of Development.   One's chronological age and level of maturity contribute to one's experience of a painful traumatic event. Forcible rape is experienced differently by a nonsexually active six-year-old, a teenager with normal sexual stirrings, a spouse with a satisfactory sexual adjustment, and an elderly woman attacked in her own apartment in congregate housing. Each year of our lives increases the range of our life experiences, fosters our understanding of the complexities and pitfalls of everyday life, shapes our capacities to cope with life events, and strengthens our abilities to bear pain and grief. As a general rule of thumb, the younger and more immature the victim, the greater the likelihood of long-term negative consequences. Multiple Personality Disorder, the most severe form of PTSD, is usually found in an adult who was severely abused as a child.

Dysfunctional Family.   If a person comes from a home that was marked by marital strife, parental alcoholism, parental substance abuse, or continuous severe verbal abuse and battering, that person will often perceive him or herself to be unwanted, neglected, and unloved. Such family chaos also precludes learning more normal ways to cope with life, so that this person's capacity to respond to traumatic events may be doubly limited. For example, the adult children of alcoholics grew up in homes with continuous unpredictable confusion because the drinking parent's brain has been chemically altered. Such children may learn to be excessively rigid in response to a need to control the family chaos. Dysfunctional families do not teach their children the good coping skills nor do they provide the support necessary for responding to traumatic events. Children who come from more normal families where they are taught the skills of stress-resistant persons have a much better prospect for full recovery.

Past History of Abuse.   As we have seen, untreated traumatic events often result in kindling, where a small amount of norepinephrine can produce intense PTSD arousal in victims who later encounter symbolic reminders in real-life events. Such a sensitive nervous system is easily more impaired, and the victim's capacity for mastery, attachment, and meaning may become less and less adaptively responsive. The research findings demonstrate that many untreated victims of assaults are more impaired in any number of health and coping dimensions, and take longer to recover. Such victims appear to be at higher risk for some of the more severe problems we shall discuss in this chapter.

Relationships to the Offender.   In a natural disaster or the sudden unexpected death of a loved one from natural causes, victims are often angry at the course of events. For most victims, however, violence freely inflicted by another human being is experienced as the more heinous and malicious act. Natural disasters may leave us feeling humbled before the power of nature, but interpersonal evil leaves us feeling raw and humiliated.

The sex of the offender may also be important. For a girl to be sexually abused by her father is despicable, but to be sexually abused by her mother may be experienced as almost beyond grief and words. A young male being battered by his mother may feel more humiliated than if he were mugged by a teenage male on the street. The importance of the sex of the assailant depends on the victim's values, and the life experiences they have had with significant others of each gender. The behavior of the assailant toward the victim at the time of the crime is often equally important. Attacks that were gentle and attentive, such as occur in some forms of incest, are sometimes easier to recover from than violence that was manipulative, exploitative, or harsh. Lastly, assailant behavior is repugnant in its own right, but if the assailant is known and trusted by the victim, the pain is usually greater, as betrayal must be added to malicious harm. To be raped or battered by one's parents or spouse is humiliating and dehumanizing. Where does one seek protection and comfort when one's parents or spouse are the enemy?

Pre-Trauma Coping Skills.   We have discussed in the second chapter the importance of mastery, attachments, and meaning as effective skills in dealing with daily life stress. Such skills enhance our capacity for solving problems, and provide us with emotional support. To the extent that individuals have these coping skills, their capacity to respond and recover from traumatic events is enhanced.

Values and Meaning.   Our sense of meaningful purpose in life helps to shape our awareness of personal significance. Such beliefs and values are important in a victim's response to traumatic events, as we have seen. The loss of a sense of a just, purposeful, and predictable environment is distressing for all victims. Combat challenges the belief that life is just and that all people are essentially good. Homicide may make life seem futile. Loss of one's meaning of a purposeful sense in life further complicates recovery if the victim's pre-trauma sense of meaning was absent or only vaguely formed.

Event Factors

Just as a person's individual life experiences shape that person's response to trauma, so does the actual event itself. The research to date has demonstrated that the impact of the event can be equally influenced by certain general properties of violent events.

Nature of the Event.   The actual type of violence may affect one's response. Violence is never pleasant, but crimes against property are usually less overwhelming than crimes against person. Being robbed at gunpoint without injury is usually less problematic than being battered or knocked unconscious. In violence against persons, being battered is sometimes easier to cope with than being raped. An episode of evil that includes both rape and battery is similarly more difficult to respond to. Natural disasters appear easier for people to respond to than interpersonal evils. Recovering from the effects of a hurricane is usually more easily accomplished than responding to permanent paralysis as a result of a gunshot from the robber sustained in a bank robbery.

Being the direct victim or being a witness to the violence is another dimension of the nature of the event. Being forced at gunpoint to watch your best friend being tortured appears less destructive than experiencing such harm yourself. Similarly, being forced to watch the rape of another is terrifying but appears less horrifying than if you are yourself raped. Witnessing the violation of others appears less destructive in most cases than being the direct victim of attack and having your own personal integrity violated.

Severity of the Event.   Severity of injuries to the victim's physical or mental health, severity of disruptions in normal living, severity of financial loss, or of standing in the community may all contribute to the victim's experience of harshness from the violent event. A car accident with minor injuries may be less distressing to some victims than a case of arson with extensive loss of life.

Victims who are blamed by the community for somehow causing the event usually experience any form of violence as more distressing than victims whose communities can acknowledge such acts as heinous behavior by others.

Frequency of the Event.   One incident of incest is unacceptable, but repeated incidents of incest have a more detrimental impact on the victim's initial response and long-term recovery. This is true for any form of violence. Repeated episodes of the same or of mixed forms of violence (e.g., a male who is battered as a child, raped as a teenager, and then wounded in active combat) make the prognosis for recovery more difficult. Each traumatic episode further impairs the victim's biological and psychological functioning, and diminishes his or her general level of resources to resume normal life. Frequent episodes harden the heart, and increase the probability of despair.

Duration of the Event.   Closely related to frequency is the duration of any one episode or of a series of episodes over time. Circumscribed, time-limited episodes are usually more easily resolved. Other things being equal, being battered weekly for one month is easier to respond to than being battered weekly for several years. Being held captive and tortured for three days at one time is psychologically less overwhelming than being held hostage and tortured intermittently over a period of five years. Over longer periods of time, pre-trauma coping skills are harder to remember and retain, and become more difficult to begin to implement again. Time can deaden the soul.

Alone/With Others.   Misery doesn't really love company, but burdens are easier to share when everyone faces it together. Painful events that befall groups of persons at one time, such as communities confronted with natural or man-made disasters, are usually more easily dealt with than when a person is the sole victim. Groups of victims have the rudiments of a support network even in their beleaguered state, and can provide some degree of emotional support and companionship, as well as the collective wisdom of the group for problem-solving in the dark days that first follow the violent event.

Environmental Factors

Environmental factors refer not only to where the event took place (a dark alley, a secluded beach), but to the response of the community of persons where the victim resides.

Safety and Protection.   Safety and protection are the first concerns. Is the victim safe from a repeat of the event? Does the individual have the necessary protection to keep the assailant at bay in the future? Removing the victim from the crime, and providing adequate police and medical attention for the victim's needs are all ways that communities can be helpful. Such interventions are crucial for protecting the victim and reducing the potential level of violence in any community. A spouse who swears out a restraining order, and then cannot get it enforced, even with repeated calls to the police, will learn that he or she is not safely protected. The assailant will learn that the community will tolerate some level of violence, and will usually seek some way to exploit this to the detriment of the community as well as the victim.

Community Resources.   A second environmental factor that can facilitate or impede a victim's recovery is neighborhood resources for help. Are there safe houses or shelters readily available? Are there victim assistance services and adequate, affordable legal aid? Are there mental health services, self-help groups, and religious counseling available for those who wish them, either during the crisis itself or in the longer period of recovery? Is there an education program on crime prevention in the schools and in the community? These are some of the basic ways that communities can protect victims and themselves from future violence.

Unfortunately, some of these resources may be a mixed blessing. For example, the very tasks that well-meaning police officers, prosecutors, and judges must engage in to provide redress for the victim may themselves increase the negative effects of the traumatic event. Cross-examination can be humiliating and may actually re-traumatize the victim as he or she must recount in specific detail the actual circumstances of the event. Such a process fosters the return of dissociated and intrusive memories.

Notwithstanding, victims recover more quickly and more fully in communities that provide the basic range of victim services.

TABLE 1

Risk Factors for Victims:

Person	Event	Environment
Biological Predisposition	Nature of the Event	Safety and Protection
Age and Stage of Development	Severity of the Event	Community Resources
Dysfunctional Family	Frequency of the Event	Community Values
Past History of Abuse	Duration of the Event	
Relationship to Offender	Alone/With Others	
Pre-Trauma Coping Skills		
Values and Meaning		

Community Values. The final environmental factor is in many ways the most important. How does the community make meaning of traumatic events as a community?—This reflects the basic societal values in that community. Does the community support its victims? A community that understands that acts such as rape and battering are crimes, that the victim did not ask for this to happen, that the victim does not have a need to be punished by such acts, and that victims are deserving of medical care and legal redress is helpful to the victim in the process of recovering. Such communities stand for basic values of human respect and justice, and can help to buffer the potential negative effects of such evil acts just as supportive families can do.

Victims who become pariahs in their families and communities become further isolated from the necessary resources for reestablishing mastery and caring attachments. Even in our own time, it is still not uncommon for a victim not to be accepted even by his or her own family. A young teenager tells her mother of the father's incestuous behavior only to find herself ordered out of the house by her mother. Some communities remain equally unaccepting. To the extent that families and communities behave in this manner, making meaningful sense of what has happened is not only delayed, but compounded in difficulty by the hostile reaction of others.

Table 1 presents a summary outline of the person x event x

environment risk factors associated with the potential negative impact of traumatic events. You should take a few minutes at this point to assess any risk factors that may have been associated with your own violent episode. Do not become discouraged if you have several of the risk factors. This will not preclude recovery, but it is important to understand the various ramifications of what has happened to you so that a full plan for recovery can be thought out.

The Disorders of Untreated Psychological Trauma

With some understanding of the risk factors that can make the medical consequences of untreated trauma worse, we now want to begin our review of these conditions to see if any of these medical conditions may be a part of your current life and your past abuse history.

In doing so we need to remember that any signs or symptoms you may have that appear to be related to abuse may be signs of other medical illnesses that are not related to abuse, even if abuse has occurred. For example, victims often have signs of anxiety or depression. Addison's Disease (not enough adrenaline) and hyperthyroidism (overactive thyroid gland) can produce symptoms of anxiety just as cancer of the lungs is often first experienced as depression. In these cases, the anxiety or depression are due to medical illnesses, and are not necessarily related only to PTSD.

For these reasons, as noted earlier, it is very important for all victims to seek medical advice to be sure that the problems experienced are correctly diagnosed. Your physician or medical specialist will help you correctly identify the specific illness, and suggest the necessary treatments and self-help groups.

While it is best to see a physician first in these matters, I realize that many victims initially feel uncomfortable about this (e.g., the female rape victim whose primary care physician is male, the battered spouse who feels possible surgery may be further disfiguring). If this is true for you, you may find it helpful to seek out some family member, friend, or counselor whom you can trust and respect. Discuss the problem in a general way, and find a process that is comfortable for you so that you can gain medical input. Counselors, especially those who work primarily with victims, can be helpful intermediaries at times like these.

The Anxiety Disorders

Given the nature of the biology of trauma, it is understandable that anxiety-based disorders have high percentages of untreated victims. We have already discussed posttraumatic stress disorder in detail, but there are other anxiety-related disorders that afflicted individuals may not link to either PTSD or the original traumatic episode when they should be doing so.

Panic. The first of these is panic disorder. While we have seen that panic may be found in PTSD, often panic is experienced by itself and without other PTSD symptoms such as intrusive memories or nightmares.

Individuals who experience the anxiety state of panic disorder often experience this terrifying fear as sudden and unrelated to anything going on at the moment. Persons in panic may experience chest pain, choking sensations, hyperventilation, dizziness, sweating, faintness, trembling, and fear of dying or going completely out of control.

This very unpleasant and frightening experience is not always related to an abuse history. Panic seems to run in some families as a genetic illness; in other persons panic disorder sometimes accompanies serious clinical depression. Still, there are many cases of panic in trauma victims that arise primarily when victims label normal bodily sensations as catastrophic (Hawton et al., 1989). The end result is panic, and, unlike victims with classic PTSD, victims who experience panic only may no longer be drawing a connection between their current panic attacks and their past abuse.

Obsessions. This is the medical term for people who are constantly worrying and this may be another pathway for expressing the aftermath of untreated trauma. When a traumatic event occurs, some victims in their hypervigilance begin to scan everything in their environment as a potential source of threat. Unlike the panic-disordered victim, obsessive victims have not made catastrophic labels about their hypervigilant bodily sensations, so they do not experience panic. They do, however, experience intense anxiety stemming from the persistent fear that some situation or some person is dangerous or may become so. Some worriers worry during the day; some worry predominantly at bedtime so that they cannot fall asleep; some seem to worry all the time. For example, children of alcoholics often worry about whether the drinking parent will be sober, whether that parent will stay sober, and what they will do if the parent is intoxicated. Victims who obsess worry when things are going poorly, to be sure; they may also worry when things are going well for fear that it is not really going well, or that the good times will soon pass into darkness. Such obsessional rumination paradoxically may help the victim to feel in control initially.

Recent medical evident has also found that some worriers are born this way. Worrying is one cognitive style by which the brain processes information, methodically going over each detail again and again. "Born worriers" who are victims may end up worrying even more.

Somatoform Disorders.   Some victims speak of their PTSD pain with their bodies. Somatoform disorders are disorders in which a person's psychological distress is expressed in bodily symptoms rather than in words, feelings, or recurring thoughts. Such persons use their bodies to express their psychological needs. These somatic illnesses can be used to communicate interpersonal conflict with another. For example, the patient may say he or she would like to go to the movies, but his or her chronic pain prevents this. Such a person is really saying that he or she does not want to go to the movies. Some somatoform disorders are utilized to solve family problems, such as in the case of a child who suddenly goes to bed ill, and in effect curtails the parents' arguing. Victims are often found among somatoformly ill patients. Some victims may develop a somatoform disorder because they are unable to give themselves permission to take care of their own needs, and being sick enables them to accept a dependent role with its offer of help from others.

These somatoform disorders may be expressed in many different types of medical problems. Some persons develop traditional psychosomatic disorders such as asthma, ulcer, irritable bowel, dermatology problems, and the like. Some persons have hypochondriasis. In such cases, the victim believes that every ache or pain reflects a serious or potentially lethal illness. Medical tests demonstrating that the person is in good health do not reduce this morbid preoccupation with the fear of illness.

Briquet's Syndrome is another type of somatoform disorder, and is found in one to two percent of all women. These patients come to the doctor with complex histories of multiple symptoms, and many hospitalizations and surgeries. Often they are abusing substances, and, like hypochondriasis patients, many have past histories of abuse.

The last somatoform disorder to concern us is chronic pain. It, too, may be found in victims with past abuse histories. Chronic pain is a continuous hurting in some part of the body. Recurring low back pain, neck and shoulder pain, and hip pain are all common examples of chronic pain states. Why some victims develop these chronic states is unclear. Perhaps the victim is attempting to communicate that life is painful and unbearable. Perhaps there is a link between the experienced pain and the depression many victims experience. Medical researchers have recently learned that many of the pathways for pain in the body are the same pathways as those for depression. It may be that the depression and the distress associated with PTSD can somehow lead to biological changes that result in chronic pain, and research is now underway to help us understand these possible links.

The Depressive States

Serious clinical depression is common in victims because of the losses they have sustained from the traumatic events, and because those events in time lead to decreases in serotonin in their body chemistry. Depression is marked by a loss of energy and a loss of interest in one's daily life and goals. Sometimes victims are also sad, tearful, angry, or preoccupied with guilt and hopelessness. Such persons may lose their appetite, have irregular sleep patterns, have a decrease in their interest in sex, and have trouble concentrating and remembering. It is as if the body comes to a slow, grinding halt. In the more severe cases, victims may think about taking their lives, or actually attempt to do so.

Some forms of depression appear to be genetic illnesses, and run in families like other genetic illnesses. Other forms of depression appear to result from severe and prolonged life stress other than traumatic abuse. Still it remains true that many of the depressed people in the world are victims of traumatic abuse who have sustained painful losses to their physical integrity and their sense of self-worth and personal significance. Many remain in a state of unresolved grief, and many have been depressed for so long that they have come to think of their current state of functioning as normal. They have habituated to their state of chronic unhappiness.

Severe depression may lead to suicide. Sadly, some victims follow this path. Suicide may result from intentional directly lethal behavior such as shooting or stabbing oneself or overdosing on pills and alcohol. It may also happen accidentally from other forms of self-mutilation such as biting, burning, head banging, and wrist cutting. Each year in the United States there are over fifty thousand suicides, and the number is growing. Each suicidal death affects the lives of at least five other persons who knew the victim.

Suicidal behavior can be divided into two types, and victims of PTSD appear to be found in both types. One form of such behavior is suicidal gesturing in an attempt to coerce or manipulate others into doing what the gesturer wishes. Another is that of people who attempt to take their own lives when a relationship such as an engagement fails. Usually such gestures are medically superficial, and the person "arranges" to be found by knowing when the ex-fiance will return or by making a well-placed phone call to announce the deed. Such gestures are meant to induce fear or guilt in the other party so that he or she will not leave. If the other party returns, the status quo before the gesture has been reestablished, and the gesturer has in effect been rewarded for coping in this manner. Because of this, there is a high likelihood that the gesturer will behave the same way when the relationship is again in trouble. Because of the risk of accidental death, this is a dangerous way to cope.

The second grouping of suicides is of people seeking relief from intense personal distress. Sometimes they feel rage or guilt, despair or hopelessness. They hope that death will result in a less painful rebirth, reunion with a deceased loved one, revenge on someone else, or serve as self-punishment for perceived wrong-doings. Such suicidal individuals have the perception (not always accurate) that no one is there to help out with these intense feelings. The suicidal act ends the sense of abandonment and its accompanying painful distress. Often when the person in such distress finally makes a firm decision to end his or her life, he or she now feels in control, and will appear fully calm and even somewhat brighter to the family or friends in the days or hours shortly before the suicide. One female victim of sexual abuse, who consistently refused treatment, bought herself a new set of clothes, took her family out to dinner, and then drove to a secluded area, where she swallowed a fatal dose of sleeping pills.

Recent medical evidence from Scandinavia (Justice, 1988) suggests that individuals in this second group may in fact have abnormally low levels of serotonin in their bodies and may have a lower threshold for tolerating life's problems. It is not known yet whether such persons are born this way or whether their serotonin is somehow used up more quickly in the face of stress. It is helpful to think of the second grouping of distress suicides as resulting from the imposition of an acute depression in a person already suffering from chronic depression. This lowered biological threshold suggests that at least some distress suicides may be true medical conditions.

The Addictions

Since we have previously outlined the various possible links between PTSD and addictive behavior, we shall focus our attention here on the specific types of addictions.

Addiction is marked by a progressive loss of control of one's life. There is loss of control over the addictive behavior itself as it becomes progressively worse as the years pass. There is estrangement from one's loved ones and friends as the addiction becomes more central in the person's life. Feelings of anger, sadness, rage, shame, and depression are present. Daily life becomes complicated as addicted persons have to rearrange their daily events to meet the needs of their habits, and there are also major changes in values as addicts begin to lie, or steal, or become verbally or physically abusive toward loved ones. The addicted individual's behavior can lead to serious health consequences for both mind and body, including problems with physical addiction. Addictive behavior is not a helpful solution to the unresolved grief of a traumatic past.

Alcohol Use Disorder.   An alcoholic is defined as someone whose drinking gets him or her into trouble. It may be trouble with the boss, trouble with the law, estrangement from spouse and/or children, or health problems. If the drinking causes him/her to get into trouble, he or she has an alcohol use problem. Note that our definition makes no reference to type of alcohol beverage, age, sex, physical strength, social class, time of day, or place of consumption. If one's use of alcohol gets one into trouble, one has an alcohol use problem. Some alcoholism appears to be genetic in origin, and some occurs as the result of excessive drinking for many years, without any history of abuse; but victims often use alcohol to sedate the arousal associated with the aftermath of PTSD, and many become addicted in this manner.

Here are four questions we use in our clinics to help people, including victims, assess their potential for abuse: (1) Do you need an eye-opener? (2) Do you get angry when people discuss alcohol? (3) Has anyone ever told you to cut back on your consumption? (4) Do you feel guilty about your drinking? An answer of "yes" to two or more of these questions indicates a problem with alcohol.

The health consequences of long-term usage of alcohol are serious. Brain dysfunction, heart disease, various types of cancer, cirrhosis of the liver, loss of sex drive, deformed infants, and serious depression (as a side effect of the alcohol itself) are all expectable outcomes. Alcohol abuse is involved in over eighty percent of all suicides.

Clearly, alcoholism is a powerful and destructive medical illness, and we are not talking about skid-row alcoholics. We are talking about persons like you and me—men and women who work, raise children, take care of their homes, and are involved in church and community activities. The difference is that the addicted person/victim is greatly impaired in his or her capacity to respond effectively to these basic life tasks.

Substance Use Disorder.   Victims also use a variety of drugs to dampen down the arousal phase of PTSD, or to increase endorphin circulation to avoid endorphin opiate-like withdrawal and feelings of depression.

TABLE 2

Substance Use and the Self-Medication Hypothesis:

Substance	Type of Psychological Distress
Amphetamines Cocaine }	Depression
Alcohol Barbiturates }	Anxiety
Opiates }	Anger

As with alcohol, victims have problems with substances if the substance is getting them into trouble in any of the ways mentioned above for alcohol. Substance users run the same risks of severe physical addiction and a variety of major medical illnesses as do alcoholics. To the extent that substance users are trauma victims engaged in self-medicating the biology of trauma, they are aggravating an already compromised biology and making recovery more problematic. These unhappy outcomes can follow from the abuse of prescribed medicines as well as illicit street drugs.

In addition to self-medicating the biological consequences of PTSD, many drug-addicted victims are also self-soothing painful psychological feeling states. One of my psychiatrist colleagues at Harvard Medical School, Edward Khantzian, M.D. (1985), has a theory about which drugs are used to soothe which painful psychological feelings. Table 2 presents his Self-Medication Hypothesis. As we can see, amphetamines (including diet pills), crack, and cocaine are used to relieve depressive feelings. Alcohol and barbiturates (sedatives) are utilized to reduce anxious states, and opiates seemed to be taken to relieve anger and rage.

Substance use in any form destroys the victim's health and well-being, and can lead to birth defects in children. It does not effectively treat the traumatic abuse issues, and, in general, precludes the learning of more adaptive ways to solve any of life's problems.

Sexual Addiction. Spending one's childhood in a dysfunctional family often leads to feelings of being worthless and unlovable. The loneliness, sadness, and depression associated with these false beliefs may lead to the utilization of sexuality as a way of self-soothing the pain.

A sexual addiction is a state in which the individual is constantly seeking sexual gratification. Self-stimulation, sexual activity with others, a continuing interest in pornography, prostitution and the like may all be signs of a sexual addition. Such persons have a voracious sexual appetite that never seems appeased. It is particularly common among Adult Children of Alcoholics.

Personality Disorders

Personality disorders are long-term, ingrained patterns of behavior that are ineffective in coping with the adult stress of everyday life. They are sometimes found in victims who come from dysfunctional families where they were both abused and not taught more effective ways to cope with life's problems. This inability to cope with adult life-stress often results in an additional series of adverse life experiences that further compound the impaired functioning of these already vulnerable persons. These faulty coping patterns are so habitual that personality disordered persons do not usually learn from their mistakes.

There are many different types of personality disorders, but, thus far, three have been identified as having a high number of trauma victims: Antisocial, Borderline, and Multiple Personality Disorder.

Antisocial Personality Disorder. This personality disorder involves a long-term pattern of aggressive behavior toward others. These individuals, usually men, have a long and continuing history of disregarding society's rules and values. Many were sexually and physically abused as children, and became truant, delinquent, and involved in lying, theft, vandalism, fighting, and casual sex. As adults this pattern of nonconcern for others is evidenced in their failure to remain productively employed, to be stable spouses and parents, to value the truth. They become impulsive and reckless, and engage in a range of criminal behaviors with no apparent remorse. As adult victims some deal with their traumatic anger and pain by seeking revenge on others.

Borderline Personality Disorder.   Borderline patients are impulsive and unpredictable individuals who have great difficulty in making and keeping relationships with others, and experience intense, uncalled-for displays of angry rage. Their chronic feelings of emptiness leave them almost frantic when they are alone, and can lead to the addictive behavior and self-mutilation that we have seen earlier in the case of Ethel.

Some medical researchers believe that Borderline patients may have been born with this tendency to become easily angered in the face of stress. It is thought that some disorder of the limbic system may be involved. In addition, most come from chaotic homes where incest, rape, and battering were common. Recent medical evidence has shown that as many as half of such Borderline persons have been sexually abused, half have been physically abused, and over sixty-five percent have witnessed serious domestic violence (Herman, Perry, & van der Kolk, 1989). Most often these Borderline persons are female. About one-third have untreated PTSD.

Multiple Personality Disorder (MPD).   In states of traumatic stress, victims may have dissociative experiences. As we have noted: the most severe dissociative state is called Multiple Personality Disorder (Braun, 1986; Putnam, 1989), and eighty percent of persons with MPD have a history of severe physical or sexual abuse in childhood. MPD includes several different "personalities," reflecting differing facets of the victim's life. There might be a "good" personality, a "mischievous" personality, and an "angry" personality. Usually one personality knows that all the others exist, and it is called the host personality. As we have seen, these "personalities" represent this extreme state of dissociation. Sometimes it is more helpful to think of these various personalities as different aspects of the consciousness of one person who has been so overwhelmed by traumatic stress that he or she cannot integrate these several aspects into the one whole personality that most of us have.

A person with multiple personalities can have anywhere from two to fifteen different personalities. Often each has its own name. This illness usually manifests itself at around thirty years of age and is most often found in women. This form of extreme dissociation is the victim's way of trying to cope with continuing bizarre behavior during childhood. Being severely strapped, being locked in a closet for long hours, going days without food, bondage, almost complete emotional neglect, in addition to the sexual abuse are all factors that may contribute to the emergence of multiple personality disorder.

* * *

We have now finished our summary of the various medical and psychological problems that may arise when victims of violent and malevolent acts are left untreated. This chapter has been expressly written so that you may determine whether or not you are suffering from any of the symptoms noted above. This review will also, hopefully, reinforce the importance of your having a proper medical evaluation as a necessary step in identifying the exact problems you will need to address for full recovery. There are excellent methods for treatment and recovery; thus, it is important to remember that it isn't over till it's over, and it isn't over till it's treated.

This chapter brings to a close our review of the basic events that occur in all victims of PTSD. Hopefully, you now have a deeper understanding of PTSD, its symptoms, and the various psychological and biological changes that may occur in its aftermath.

In the next section of the book we shall explore specific traumatic stress syndromes that can lead to these PTSD responses in any of us. Because our knowledge is still limited, we shall include only those traumatic events where there is a reasonable amount of information for us to have a basic understanding of that particular form of violence. Included in this overview will be sexual abuse, physical abuse, combat, and family alcoholism.

As you read these next three chapters, consider each syndrome from the PTSD perspective that we have just outlined. In each case consider how mastery, attachment, and meaning are disrupted and how body chemistry may be altered. It is syndromes such as these which, if they are left untreated, may result in the serious illnesses that we discussed in this chapter.

For those of you who wish to study the scientific evidence in greater detail, the Select Readings for each of the next three chapters have the recent references for each particular syndrome.

Part 2

POSTTRAUMATIC STRESS DISORDER: THE VICTIMS

5

Sexual Abuse

There is a little less sunshine in the world today.
— Beverly Sills

We will meet your physical force with soul force.
— Martin Luther King, Jr.

Susan thought he was the most handsome guy she had ever seen. Tall, good-looking. Halfback on the football team. One of the better students in the junior class. And he had asked her out.

She, the college freshman, was going to the team barbecue at the beach on Saturday night. For a passing moment she had been uneasy about this party, but excitement overtook her. She'd have to get her hair done, new clothes, new shoes. She had been over the details of this date at least one hundred times with her roommate Sheila. Wally had even sent her a rose. He thought of everything.

He certainly did. Susan was his mark. He had planned every last detail—even the rose. Slim-figured, pleasant-mannered Susan was his choice because she was so very trusting. He smiled to himself as he quaffed his beer. Stalking the prey was half the fun.

He picked her up at six-thirty that evening. Susan had decided to wear to the beach the pearl necklace that her mother and her grandmother had each worn on their wedding days. This too was a special time, and she looked striking.

They were to go with two other couples, and drove to the dorm to get them. His teammates said their dates were running late so

Wally poured everyone some scotch that he had purchased earlier. This was the first time Susan had ever tasted scotch and the odor made her sick, but she carried on valiantly. After all, this was college. Wally's teammates suggested that Wally and Susan go on ahead, and join the rest of the team at the beach. They would join them later. Wally and Susan had a second one for the road.

Wally chose a soft-rock station as they drove the ten miles to the lake. Susan asked him about his family, his career plans, and his future as they drove toward the setting September sun.

When he stopped the car, Susan saw no barbecue and no team. Somewhat light-headed from the alcohol, Susan froze now in alert terror. "There is no team, there is no barbecue, there is only the two of us." He pulled a knife and ordered her to disrobe. He threw her to the ground and pinned her with his weight, and raped this virgin freshman three times at knife-point. "You tell, you die." Then he was gone.

Susan sat there in the sand. Bruised, unable to speak, crying softly for her mother, she was stunned and repulsed by her own body. It had been so dirty, so ugly, so violent. Her dreams of marriage until death were now but ashes in her heart, and she felt like a woman of the streets. She fumbled for the pearls at her neck.

It was dark.

*   *   *

Sexual abuse has been a part of the human family's inhumanity toward others for thousands of years (Brownmiller, 1975). Rape is listed as one of the four major violent crimes, along with murder, assault, and burglary; and Susan is one of its current victims in a form of sexual abuse known as date rape. Sexual abuse may occur at any age, for either gender, and across all social classes. It is the one violent crime in which women are more likely to be victims than men.

Accurate estimates of rape are hard to ascertain because of the various reporting problems that we mentioned in chapter 1, and because of differences in the way sexual abuse is defined. Many jurisdictions follow a legal definition of rape that is defined as forced penetration without consent. Some in the women's movement and some social scientists speak of the experience of rape which is best understood as any unwanted sexual behavior with a nonconsenting adult. Forced undressing, petting, touching, masturbation of others, oral, anal, or vaginal intercourse, bestiality, and forced voyeurism of others engaged in sexual activity might all be subsumed in this understanding of rape.

Some estimates report that one woman in four and one man in five will be sexually abused in her or his lifetime. For the reasons we have listed, however, it remains difficult to accurately gauge the extent of such abuse, and more research is needed. Even so, most everyone agrees that there are more episodes of sexual abuse in our society than any of us would want.

In this chapter we turn our attention first to the general nature of sexual abuse, and then we will review the specific types of abuse that adults and children encounter. Even though such abuse can be successfully treated, none of it is pleasant. If you are a victim of such abuse, you know that to be painfuly true. If you are not a victim of sexual abuse or have not worked with victims and their issues, what you are about to read may be truly astonishing to you.

The General Nature of Sexual Abuse

Sexual abuse is an act of domination over another, but there is no single way to explain why sexually abusive persons behave this way. These crimes are complex forms of behavior, and scientists are trying to understand why sexuality and aggression become disorganized in some men and women. Here is what the research findings have yielded thus far.

In males, some abusers are seeking sexual gratification and resort to force to obtain it. Some are primarily seeking to inflict harm, humiliation, or degradation, and use sexual abuse as the method to attain these ends. Other male abusers are self-centered and selfish. They want their own way or are seeking revenge for some perceived injustice (Gelinas, 1979; Herman, 1981, 1992). Still other males abuse others to give themselves the illusion of power, or to keep women in social, political, or economic subjugation. Some are outright antisocial persons for whom criminal behavior is a way of life.

Sexual abuse perpetrated by women appears much less in extent,

and has been less studied. A few women do appear to use physical force to attain sexual gratification (Sarrell and Masters, 1982).
(Men and boys are capable of becoming sexually aroused even in
states of high anxiety or fear.) Some abuse by women occurs in
baby-sitting situations, and some, of course, is incestuous. Other
female assailants appear primarily motivated by the need for the
illusion of power and control as do some male assailants. Still
others seek contact with others, and some seek to forge a sense of
belonging with males by enticing innocent women into sexually
dangerous situations such as fraternity gang rapes (Sandy, 1990).

Myths of Rape.   There are false beliefs about rape that sanction
such behavior. These beliefs imply that victims of rape really want
to be victims of this crime.

Writer Robin Warshaw (1988) has outlined some of the more
common rape myths about women. These include the belief that
rape is spontaneous, usually committed by strangers, and often
set in motion by uncontainable sexual arousal from kissing or
touching. Some believe that only virgins can be raped unless a
lethal weapon is present. Victims, as well as assailants, often believe that if the victim experienced physical pleasure or did not
fight back, then no rape occurred. All of these common cultural
myths are false.

Why, then, are some women raped and not others? There are
three common theories that have attempted to account for why
the violence happens to some women and not others. All three
theories, like the myths, imply some fault in the victim. The first
theory states that women who have a past history of abuse are at
higher risk for subsequent repeat victimization. The second theory
suggests that some women are at greater risk because they are less
dominant and less socially aware. A third view proposes that certain situations can increase the risk of rape. Included here may be
a pattern of behaving recklessly or being known by the neighbors
as a bad apple. Which theory of women victims is correct? A
recent study found that all three theories were not particularly
good at predicting who would become a victim in the majority of
cases. Most crime is a function of chance: the wrong person in
the wrong place at the wrong time.

What of men who are raped? We know remarkably little (Lew,
1988). I have had male patients tell me twenty, thirty, even forty

years later that I was the first person they ever told of their abuse. Other counselors have had similar experiences with male patients. This is a secret trauma for men as well as women. Men can be raped at home as well as in open areas. Additionally, some are sexually attacked when they are incarcerated. Some male victims are antisocial persons, some are addicted to alcohol and drugs, but most are ordinary people to whom no blame is attached. Since men are expected to be stoic in the face of adversity and do not report such humiliation, there is even less known about male victims of rape than about women victims.

As a society, we need to stop blaming the victims. The myths are not true. No theory of selection is true. No man or woman asks to be raped. No person wants to be a victim of violence.

Robin Warshaw (1988, p.52) has compiled a summary of the characteristics associated with men who rape, and these findings may be found in table 1. This list does not mean that people with these characteristics will necessarily become rapists, but it does provide us with some early warning signs. While this list was compiled for women, it can equally serve as a useful guide for males who can also become prey.

The Stages of Rape. Every female victim of rape responds to the trauma in a series of stages (Koss and Harvey, 1991). The stages are remarkably uniform whether the event is date rape, incest, rape in marriage, or of any other type. There is no corresponding research on male victims, but it is reasonable to assume that they go through a similar process.

The first stage is the acute crisis phase. For example, Susan in our chapter vignette, when she realized she was about to be attacked, froze in alert attention. The acute crisis phase involves the mobilization of the body's and the mind's emergency response systems. Epinephrine prepares the body for attack, and norepinephrine and endorphins place the mind in full alert so that survival is enhanced. As we have seen, the victim may be shocked or stunned or frozen in terror. Denial or dissociation may occur as the victim seeks to distance herself from the event psychologically and as she decides the best strategy of escape or the best strategy of coping effectively to minimize the potential negative effects of the violence. When the event has immediately passed, victims become highly anxious and sometimes frantic in these

TABLE 1

Some Characteristics of Potential Rapists:

Emotionally abuses you (through insults, belittling comments, ignoring your opinion, or by acting sulky or angry when you initiate an action or idea)

Tells you who you may be friends with, how you should dress, or tries to control other elements of your life or relationship (he insists on picking the movie you'll see, the restaurant where you'll eat, and so on)

Talks negatively about women in general

Gets jealous when there's no reason

Drinks heavily or uses drugs to get you intoxicated

Berates you for not wanting to get drunk, get high, have sex, or go with him to an isolated or personal place (his room, your apartment, or the like)

Refuses to let you share any of the expenses of a date and gets angry when you offer to pay

Is physically violent to you or others, even if it's "just" grabbing and pushing to get his way

Acts in an intimidating way toward you (sits too close, uses body to block your way, speaks as if he knows you much better than he does, touches you when you tell him not to)

Is unable to handle sexual and emotional frustrations without becoming angry

Doesn't view you as an equal, either because he's older or because he sees himself as smarter or socially superior

Has a fascination with weapons

Enjoys being cruel to animals, children, or people he can bully first efforts to cope. Am I injured? Do I need medical care? Will I become pregnant? Whom should I tell? Should I go to the police? Will my family understand? Was my judgment that poor? I feel dirty and defiled. This first stage can last from a few hours to several days. Victims will remain highly anxious during this period, and may develop the symptoms of PTSD. They are seeking some degree of initial mastery over the immediate confusion in their life, and a safe place to be.

The second stage deals with the return to some degree of the victims' usual daily routines. Although victims may continue to be shaken by the event and may continue to blame themselves as a way of maintaining self-control, the intense anxiety abates and attention is focused on getting back to normal. Returning to work, school, child-care responsibilities, community projects, and the like occupy the victims' energies. The anger and fear toward the assailant is temporarily put aside, and the return to a more normal routine calms and reassures victims that they have emerged less ruined than they first believed. Loved ones can be immensely helpful during this period in aiding the restoration of more normal routines. Reasonable mastery and some degree of caring attachments are reestablished.

The third stage may come weeks or even years after the rape. The immediate crisis has passed, normal daily lives have been restored, and victims now have time to reflect on the event. Victims usually become depressed at this point. They realize that their lives have been dramatically altered, and they begin the search to understand what has happened and to make some meaningful sense of the violence. This is often done by talking to caring others or to other victims, by reading everything that has been written on the subject, seeking professional counseling, or discussing these evils with the clergy. In time, the victim can make some meaningful sense of what happened and can establish a new, or restored, more mature meaningful purpose in life.

Victims who do not complete the tasks of these stages are at increased risk to enter the chronic phase of PTSD, and to become chronically depressed.

The Impact on the Victim. These malicious acts by twisted human minds may affect victims in a range of ways. Health is the first area of impact. In addition to the possibility of medical injury, general discomfort, and possible pregnancy for females, each victim has entered a state of high physiological arousal. Many victims will experience the physical and intrusive symptoms of PTSD, and some will seek to withdraw and remain avoidant. Any of the problems we outlined in chapter 4 may develop over time in untreated victims.

Similarly, traumatic events can stunt or disrupt longer-term general functioning in basic life activities. This is again especially true of untreated victims. Formal schooling, satisfactory marriage, child-rearing, level of career promise, and general involvement in community life have all been known to be adversely affected.

A third area of impact is in the domain of feelings. Victims continue to experience a general level of fear as well as specific fears of death and of the assailant returning to repeat the crime. Experiencing anger is difficult for many victims because it is associated with being killed by the assailant. Thus anger and rage may alternate with feelings of helplessness and vulnerability in the face of life's many potential onslaughts. Victims experience sadness and depression over the loss. Many feel guilty and shameful. In some this may be self-blame to regain control. In others there may be the genuine belief that they are responsible because they complied with every directive of the assailant.

This last phenomenon is known as the "Stockholm syndrome" or survival bonding, a process first noted in 1973 in a Stockholm bank robbery. Research has shown that in time of intense fear some victims cooperate and sometimes even come to admire the assailant. They do exactly what they are told and try to emulate the assailant. This appears to be an unconscious method to ensure survival, a variant of trying to join those whom you cannot defeat. Victims are later horrified that they behaved this way and feel guilty, but it is a relatively common and automatic way in which some victims respond. This phenomenon has been noted in battered women, incest victims, hostages, abused children, and concentration-camp survivors.

The last area of potential major impact is in the area of making some existentially meaningful sense of life. We have noted Becker's (1973) discussion of the importance of finding a transcendent meaning in life, and that this is best accomplished by loving others and being concerned for their welfare. In our culture, one's sexuality is the most personal, private, biological way to express love for others. Marriage, giving birth, giving and receiving intense personal pleasure, warmth, and comfort are all ways of loving others through one's freely chosen sexual expressions. Rape temporarily destroys free choice and temporarily shatters sexuality as a mode of transcendent meaning. The unwanted penetration of one's body is an invasion of the mind and of the heart and of the soul. And the scars of this crime endure if it is left untreated.

Types of Adult Sexual Abuse

Date Rape

Most rapes do not occur when some stranger jumps out from behind the hedges and rapes an unsuspecting person. Fully seventy to eighty percent of all rapes are acquaintance rapes in which the victim knows the assailant personally. Incest, spousal rape, and date rape are all examples of acquaintanceship rapes.

It has been estimated that as many as twenty-five percent of all college females have been victims of date rape, and many do not realize that they are victims. This chapter's vignette is one clear example of how date rape may occur in some cases. The assailant, Wally, had selected his prey, had carefully planned the attack, had plied the victim with alcohol to impair her judgment and coordination, and had managed for them to be in a secluded location far from help.

The average age of the perpetrator and the victim in date rapes is about eighteen to nineteen years of age. Usually, the coercive male accepts interpersonal aggression as a way to solve problems, often has a stereotypic view of the sex roles, and wants to be approved by his peers. Sometimes the assailant also comes from a family with a past history of violence against its own members. The female victim is usually less assertive, is openly trusting, and wants very much to be accepted. Both the victim and the assailant are usually intoxicated. Much date rape goes unreported because the victim is unsure that she has been raped, because she wants to protect her family, or because she is unaware of what to do and what community resources are available.

What should a victim do at the time of the attack itself? There is no general common strategy for all cases. Some victims may see the opportunity to act quickly and run away. Some may yell for help. Others may keep the assailant talking to buy some time. The presence of a potentially lethal weapon complicates such strategies. Each victim needs to select the strategy that seems most suitable to a particular set of circumstances.

Fraternity Gang Rape

A variant of date rape for college students is gang rape which occurs in fraternity houses on college campuses. Such rapes are often referred to as "pulling train" because one man follows another. Female victims who go to a fraternity party to seek warmth, belonging, and acceptance are lured by one frat member to go off and have sexual relations with him. Often such victims are then forced to have sex with anywhere from two to eleven other fraternity members during the same encounter (Sanday, 1990).

This form of rape is again aggression in a sexual guise, and it is the way some men express their masculinity, exercise their sense of power, and receive peer support. This is a type of sexual abuse in which women are often accessories, as the older women scout the campus for younger female victims. In this manner, the older women ensure their acceptance by the male fraternity, and receive special status and consideration (Sanday, 1990).

Fraternity gang rape has negative consequences for society as a whole. Such acts degrade women, and thus degrade the healthy dependency a male must have on the female to have his sexual desires satisfied. Degrading sexual acts separate feelings of love and compassion from sexual behavior. Such callousness may contribute to marital difficulty, indifference to child-rearing, and a general attitude of limited concern for the welfare of others.

Fraternity gang rape, like date rape, is a crime and should be addressed as such by the college community.

Rape in Marriage

One in seven women is estimated to be raped by her own spouse in her own marriage bed. As with date rape, victims often do not realize that they are victims of a crime. Rape in marriage is defined as the forcing of any sexual intimacy on one nonconsenting person by another person. Several behavioral scientists (Russell,

1990; Walker, 1979, 1984) have written on this topic and the list of such sexual abuse practices by one's own mate is both frightening and somber. This is especially true when we consider that one's home-life is supposed to be a refuge from the weariness of the world.

Women have been forced to have nonconsenting oral and anal intercourse. They have been forced to participate in sado-masochistic events and to have group sex against their will. Some have been forced to have sex with animals. Some have had foreign objects inserted in their vaginas, such as broom handles and wire brushes. Such sexual acts are often accompanied by pinching, biting, beating, choking, cigarette burns, or the use of cattle prods and weapons. Verbal abuse from the assailant is almost always present as well, for those women who are both sexually and physically abused.

Some authors (Herman, 1992; Barry, 1979) have noted how such assailant behavior is really a form of torture. Each of these degrading acts by the spouse is intended to break the will of the victim for the assailant's own gain and complete dominance.

Violent rape in marriage is found in all social classes. The husband is usually the assailant, and is usually intoxicated. These violent spouses usually feel that they should be waited on hand and foot, and expect their needs and wishes to be met in every instance. Sadly, this violence and tyranny often masks an overwhelming sense of powerlessness in other aspects of the assailant's life. Such desperate sexual attempts as those noted above are committed by frightened people as an attempt to establish some sense of personal meaning.

Why would a woman marry such a person or stay in such a relationship? The answers are complex, and not easily understood. Some women marry into these relationships because they do not know any better. Others have a past history of abuse and are revictimizing themselves. Others feel responsible for the man's needs and blame themselves for not adequately having measured up. Some enter and stay in these relationships because they feel they can predict when the mate will become violent and this gives the victim some sense of control.

Leaving is not as easy as it might first appear because the victim's options are often limited. Since most victims are female, they are fearful of violence outside the home, on the streets, and in shelters. They often fear the assailant will come after them or the children, and this does happen. Sometimes victims have limited economic resources and, in fact, have no other place to go. Women cope as best they can by calling the police, getting the husband sober, going to shelters, or becoming violent themselves.

Better law enforcement and more shelters for battered women have provided some assistance, but much remains to be done.

Sexual Abuse of Children

Incest

Incest is the rape or other form of sexual abuse of one's own child, of one's stepchild, or of one's relative. My colleague, psychiatrist Judith Herman (1981) has provided us with a helpful definition. She understands incest to be the corruption of parental (or adult) love in which the powerless child must pay with his or her body for affection and care that should be freely given. Young girls are thought to constitute about eighty percent of all such cases. Usually the victim is between four and twelve years of age, although there have been reported cases of incestual abuse of infants. The abuse usually occurs more than once, and most commonly stops when the victim is about age fifteen. It stops then because the victim discloses or threatens to disclose what is happening, or runs away to the streets.

Incest can be found in many types of adult-child relationships. Biological fathers sexually abuse their daughters. Mothers sexually abuse their sons and daughters. Brothers abuse their sisters; uncles, their nieces and nephews; grandfathers, their granddaughters; cousins, their younger cousins.

The mother's remarrying does not necessarily solve the problem either, because some stepfathers have been found to be even more abusive than the biological father. The reasons for this are unclear. It may be that stepfathers are men who have difficulty with adult females and marry because they feel more comfortable with children. The bonding between stepfather and stepdaughter may be weaker, and some stepfathers may be pedophiles who prey upon young children in a variety of settings. We really do not know as yet.

Family Dynamics. What characterizes the homes in which such abuse occurs? How do they differ from more normal families? The medical and psychological research reveals that many of the fathers in these homes feel very entitled to have their every need met, yet there are also reports of other incestuous fathers who are introverted and socially isolated. Part of the problem lies in the fact that many families keep such matters secret so that researchers do not know if they have a truly representative sample of such families. While there may be many family paths toward incest, psychologist Denise Gelinas (1979) has provided us with an explanation of one common pathway that seems to occur in many homes.

In some families, the father feels entitled to be waited on. This includes having household chores done by others as well as being provided with sexual gratification. Such a father is often tyrannical and frequently alcoholic. He often comes from a home where he was himself physically abused, or where there was maternal deprivation due to illness or death. He is basically very much afraid of being abandoned and unloved, as he felt he was in his own childhood.

Mothers in such families often have an abuse history themselves. Many of them were additionally "parentified" as children. A "parentified" child is one who is required to be an adult parent to one's own parents and siblings by raising children, doing homework, balancing the family budget, and so forth. As adults, their caretaking skills will be excellent, but they will also feel lonely and unloved because of the role reversal in their own families.

Such "parentified" women are drawn to men with caretaking needs. Each party hopes to receive the love and loyalty they did not have in childhood. Usually, such situations are tenable until the birth of the first child.

When the child is born, the woman feels overwhelmed by yet more caretaking responsibilities. Feeling tired, angry, and bereft of support, she psychologically withdraws. As she withdraws, the husband fears being unloved again and becomes more demanding. This vicious cycle continues and becomes even worse if there are more children. It finally ends in a state of marital discord and sexual unfulfillment.

Gradually, the father turns his attention, and then his demands, on one of his female children. The child may at first welcome the physical warmth of the closeness, but becomes increasingly frightened of the escalating demands for sexual involvement. In time, the father sexually abuses the daughter, who is expected to fill in with domestic chores as well as sexual favors in lieu of the overwhelmed mother. Often the father may be having sex with the mother as well.

Fathers ensure such cooperative behavior by treating the sexually abused child with special favors or with threats to harm the child or the mother, if the child ever says anything about such horrendous acts. The child has no place to go and no one to tell. To survive, the child must sacrifice her body and her self-esteem for protection from the elements and for assurances of being valued. Because she is responsible for the household, she herself becomes "parentified" as her mother had been before her.

The child correctly perceives that she is unloved, and goes in search of fulfilling her own dreams of being happily married in a new life away from the ugliness of the incest. Sadly, her skills as a "parentified" young adult draw her to men with high caretaking needs, and the intergenerational cycle of incest often begins again.

Impact on the Victims. Regardless of the type of incestuous relationship to the offending adult, the impact on the victim may be serious and debilitating. In addition to possible PTSD symptoms and substance use, adult women who are incest victims often have more problems with helplessness, suicidal thoughts, and depression. Caring attachments, sexual adjustment, trust, and intimacy are all areas of concern for many adult incest victims.

Male victims of incest fare no better. They too have problems with depression, suicide, and substance use. In addition, some turn to crime. Adult males are left with confusion about their roles as male adults, confusion about their sexual identity, and some are sexually predatory toward other young males (Lew, 1988).

Victims of both sexes may end up running away to a life of prostitution.

Child Exploitation

Pedophiles are people outside of the family who sexually molest children. As common as this problem appears to be, less research has been directed to these matters.

Child molesters are usually male, often have a past history of being abused themselves, and are most commonly attracted to young boys. Sometimes they act singly, and sometimes as part of a recruiting and distribution process for pornography and prostitution.

In solo cases, the adult male often entices children between the ages of ten and twelve with promises of alcohol or money. Sometimes he is a respected member of the adult community (e.g., a scout leader, clergyman, or child counselor), and can easily gain the trust of the boys. A common practice is to ply the children with drugs or alcohol, have them engage in oral, anal, and genital sex with each other, take Polaroid photographs of this behavior, and then select one young boy for the molester's own sexual gratification.

Unfortunately, children rarely report that this is happening to them. Sometimes they are blaming themselves for what is happening. Sometimes they are financially dependent on the perpetrator. Sometimes they need the affection and attention. Often they fear censure in the eyes of the community or assailant retaliation.

Rape of Children by Children

Recent years have witnessed the rise of a new phenomenon: pre-adolescent children sexually abusing other children. The rate of arrests for rape by males under age twelve has tripled in the past two decades. Rape arrests for thirteen- to fourteen-year-olds have more than doubled in the past decade. These acts are planned, purposeful aggression toward other children who are perceived as more vulnerable and less able to protect themselves.

Since young children are not born with the knowledge of how to commit rape and are not taught it in our schools, they appear to be learning it by being sexually victimized themselves or by witnessing sexual abuse either at home or in the neighborhood. This kind of sexual abuse at the hands of other children carries the same negative consequences for the victims as other forms of sexual abuse, and at the moment, there are very limited resources to treat such victims in our country.

Forced Sexual Slavery

Some teenagers or young adults are forced into prostitution or sexual slavery (Barry, 1979). Pimps, employment agencies, modeling agencies, "wives" magazines, and outright kidnapping are used to recruit unsuspecting victims who are forced to provide sexual services to businessmen, sailors, military personnel,and even immigrant laborers in some countries. These young women are expected to have as many as sixty to eighty encounters per day, and are kept against their will by threats of violence and death, as well as by the lack of any true options for escape. This practice is not well known, but cases have been reported in our own country, Canada, and many other countries throughout the world.

Prostitution

It should not surprise us that it has been estimated that at least eighty percent of all prostitutes have been victims of physical or sexual abuse. Many are also Adult Children of Alcoholics, and run away from the abuse at home only to find an equally bleak situation in the streets.

Prostitutes are raped, robbed, forced into perversions by their customers, and are often disciplined by their pimps by means of severe beatings. It is a life of human misery, degradation, and loneliness. No one enjoys having his or her body sold as a commodity or made vulnerable to possible physical injury, communicable and life-threatening diseases, and endless hours of depression.

Why would anyone remain a prostitute? Some may not be able to leave because there is no place to go, and to go home would only bring more violence. Some did not do well in school because of their childhood abuse and family problems, and prostitution may be one of the limited ways they have of earning a living. Others are addicted to drugs and alcohol to blur the memories, and work in the trade to support their habits. In truth, many prostitutes have very limited alternatives.

*   *   *

This chapter has catalogued a very painful aspect of human life. With each act of sexual abuse there is a little less sunshine in the world. Love is central in the lives of all of us and the twisted pathways noted here are a great injustice to the victims, who are good people and have been betrayed by others in such a fundamental way. It is testimony to the human spirit that the impact of sexual abuse with its many broken dreams can be overcome. Victims do forge new and trustworthy bonds with others, their depression passes, and a new sense of perspective and meaning restores the brightness to their lives.

6

Physical Abuse

Aggression is an inept attempt
at self-affirmation.
— Ernest Becker

Darkness shall not frighten us
nor distress wear us out.
— Alfred Delp, SJ

Oklahoma City was a long way from Three-Rivers, Quebec. Margarite was thinking of her childhood. The grey stone houses with their brightly colored trims. The snowy winters. And the birthdays. She remembered the birthdays most of all. It was her family's custom to celebrate each child's birthday with a visit to the shrine to ask God to protect them, each and every one; and then to go to Henri's Bakery to select that special birthday cake. Cakes in all sizes and shapes. Double chocolate was her favorite.

Joseph Gatineaux. How excited was her shy heart when he asked her to marry him. She had spent one whole day repeatedly writing Mrs. Joseph Gatineaux to see how it would feel. He was from a good family, and he was a hard worker who preferred to do things his way. He asked her to come and follow him to Oklahoma where the economic opportunities were brighter. She was the envy of every girl at Three Rivers High.

Margarite sat in the chair in the living room. Her whole body ached tonight, in particular the left side of her body throbbed. Joseph had pummeled her again. First, he beat her with her own hair brush; then, with his fists. Finally, he threw her against the wall where she landed on her left side. Now he lay asleep. His fury spent, he was snoring as if he hadn't a care in the world.

She had tried to make some sense of it all in their twenty-seven years of marriage. Over the years, his demands and jealousy had become tyrannical. It had started on their wedding night: he became angry and slapped her. She was startled and frightened, but his tearful apology softened her fear. She too had found the ceremony stressful. It wouldn't happen again. He had promised.

But it did happen again. Many times. More times then she could count. She could sense it coming. He would come home drunk, slap and kick her repeatedly, and then punch her as if she was some rabid animal to be fended off. After the first few years, the tearful apologies stopped and so did her love for the man. She stayed with him because of the children and because he said he would kill her if she ran away. And she knew he would do it.

She sat in the darkness of the night. Today was her forty-fifth birthday. The card from her in-laws in Quebec read, "Many Happy Returns on This Day."

* * *

Physical abuse, also known as assault and battering, is another of the four major crimes in our society along with murder, robbery, and rape.

Men are much more likely to be physically abusive to others than women are, and they are also much more likely to be the recipients of such abuse. Evidence of this may be found in school yards, work sites, bars, and on the streets. Physical abuse by females is much more likely to occur within the family context.

This chapter focuses on domestic violence, the physical abuse that occurs within the home. Abuse that is perpetrated by both men and women, and the young and the old, as the physically stronger prey upon the weak. Margarite's experience is, sadly, not uncommon in many homes. Better hidden in some than in others, but not uncommon. As recently as the 1970s, it was legal to beat one's spouse in many states. As a society, we are reluctant to address such issues because of our culturally held beliefs that idealize the family, that assume that a man's home is his castle, and that children should be seen and not heard.

Violence at home is a terrifying and painful reality for many families. We shall first address the general nature of physical abuse, with particular attention to what is called the *cycle of violence.* Our focus will then turn to spousal abuse, and conclude with a discussion of child abuse, the intentional infliction of harm on one's own children.

If you are in a caring relationship in which you have been physically abused more than once—even if the interval between abuse periods has been as long as ten years—pay particular attention to the description of the cycle of violence. As the research evidence accumulates, it is becoming more apparent that violence in relationships may not be as random as it might at first seem.

The General Nature of Domestic Physical Abuse

Physical abuse may be understood as the intentional infliction of physical harm on another person without his or her consent in any circumstance other than self-defense. The victim is usually subjected to repeated attempts at coercion by the assailant so that the victim will do what the assailant wishes without any regard for his or her own needs and rights. Severe verbal abuse often accompanies physical violence.

As was the case with sexual abuse, the statistics are gruesome, and equally as hard to accurately gauge. It is estimated that as many as twenty-five percent of all women will be battered at some point in their lives, that twenty percent of all women will be battered and raped, that forty percent of all pregnant women will be battered, and that in one marriage in every seven both spouses batter each other.

It is not infrequent to hear of men who are battered by their spouses as these spouses erupt in violence in the face of personal and family pressures. Since the abuse is humiliating, men report little of it, and thus, as yet, we have no accurate way to estimate its extent.

Children and the elderly fare no better. Most of us are dismayed and saddened when we read in the paper of an infant found dead in a trash bin. What most of us do not realize is that this is not such a rare event. In 1985, over five thousand children died as a result of abuse or neglect. This is over twelve children each day in our country. Half of these victims were less than one year old.

In that same year, an additional two million cases of nonlethal child abuse were reported. Moreover, estimated statistics suggest that as many as five percent of our elderly are abused (often by their own adult children). Elder abuse can include physical assault, financial exploitation, or emotional neglect.

If the statistics cited here were for any other health problem, it would be called a major national epidemic. The women's movement has made some legal and counseling gains for female battered victims, but there is little in terms of support for male abuse victims. Social service agencies are overwhelmed with the cases of children, and often only the most severe cases can be investigated. Much remains to be done.

Who batters? In over ninety-five percent of the cases the assailant is male. Such men usually feel powerless, jealous, and insecure in the face of life stress. Since victims are often less assertive because of traditional values, depression, or poor self-esteem, the aggressor uses intimidation and violence to get his needs met. This behavior is effective, and leads to escalating violence over time.

Recent research has begun to document increasingly aggressive behavior in female assailants. A recent study reported that in courtship and marriage women engage in more pushing, shoving, grabbing, and slapping than their male fiances. Although such behavior is less injurious, there are additional reports of a few women committing crimes of a more violent nature. It is too soon to know whether such behavior in women is generally on the increase.

There are several theories of violence (see Walker, 1984, for a review). Sometimes the assailant does not have adequate interpersonal skills for developing caring attachments and effective problem-solving. Factors such as unemployment and general economic stress, poverty, too many children, poor health, childhood trauma, and alcohol and substance use (particularly crack/cocaine) have also all been associated with violent behavior. Whatever the reason is in specific situations, one factor is common to all forms of physical abuse: if left unaddressed, over time the abuse itself will become progressively worse. Unfortunately, it is often the only thing the victim can count on.

Myths of Physical Abuse. Just as our society has false beliefs about rape, it has similar faulty notions about physical abuse and battering. Lenore Walker (1979, 1984), an expert on the battering of women, has outlined some of the basic myths about battering that we know are not true. It is not true that battering occurs in only a small percentage of our fellow citizens, and it is equally not true that it occurs only in working class, minority, or poor families. Battering does not stop when people get married. Battered women are not crazy, they are not asking to be mistreated, and they cannot always just leave the situation for reasons we shall mention shortly.

There are equally false beliefs about assailants. It is not true that batterers are a few antisocial persons. It is not true that all batterers are unsuccessful in general and lack the skills to cope in society, and it is not true that such men are violent in all of their relationships. (This is true in only about twenty percent of the cases.) More common is a pattern of violence toward selected members of the family.

It is also not true that batterers cannot be loving partners, or that such relationships cannot change and improve. Finally, increasing medical evidence in the field of traumatic stress suggests that staying in a battering relationship for the sake of the children, when there is no effort on the part of the batterer to stop, places the children at increased risk. It is not in their best interest. The children are at risk to be battered themselves (one-third of the cases), and to become batterers in their adult lives as a consequence of witnessing spousal traumatic abuse.

Dr. Walker (1979, p.254) has compiled a list of traits associated with potential male batterers. Table 1 presents these features. Many of these characteristics would also be true of female batterers. As with the markers for potential rapists, these signs for batterers do not mean that any specific individual will become a physically abusive person, only that the risk for such events is increased.

The Cycle of Violence.   Several social scientists have paid particular attention to spousal abuse, and their research has greatly advanced our understanding of domestic violence and physical abuse in general. Dr. Lenore Walker (1979; 1984) was the first to report on a three-stage process of violence in the home, and she refers to this process as the cycle of violence. Here is what she has to teach us.

The first phase is referred to as the tension-building phase.

TABLE 1

Some Characteristics of Potential Batterers:

1. Does a man report having been physically or psychologically abused as a child?
2. Was the man's mother battered by his father?
3. Has the man been known to display violence against other people?
4. Does he play with guns and use them to protect himself against other people?
5. Does he lose his temper frequently and more easily than seems necessary?
6. Does he commit acts of violence against objects and things rather than people?
7. Does he drink alcohol excessively?
8. Does he display an unusual amount of jealousy when you are not with him? Is he jealous of significant other people in your life?
9. Does he expect you to spend all of your free time with him or to keep him informed of your whereabouts?
10. Does he become enraged when you do not listen to his advice?
11. Does he appear to have a dual personality?
12. Is there a sense of overkill in his cruelty or in his kindness?
13. Do you get a sense of fear when he becomes angry with you? Does not making him angry become an important part of your behavior?
14. Does he have rigid ideas of what people should do that are determined by male or female sex-role stereotypes?
15. Do you think or feel you are being battered? If so, the probability is high that you are a battered woman and should seek help immediately.

During this time interval, the spousal victim can sense that the battering spouse is getting increasingly frustrated, short-tempered, hostile, and agitated. The victim intuitively realizes that a violent episode will follow.

These tension build-ups are cyclical in nature. Unlike street crime where the battering is sudden, impulsive, and soon over, violence in long-term relationships has a predictable cycle. It may be each day, every two weeks, every two years, once in five years—but there is a cycle which can be predicted if the victim keeps track of the occurrence of such events. Sometimes there are anniversary events that are associated with the outburst (e.g., anniversary of the spouse's mother's death); more often the precipitants are not as clear, but the timing of the cycle can often be reasonably estimated.

In the second phase, the inevitable violent outburst occurs. The victim is threatened, attacked, even stalked and mentally tortured before the battering. The assailant loses control, and fury is unleashed. The violence can last from a few minutes or hours to a week before the batterer stops. Other than the obvious emotional and physical exhaustion, the reasons why the assailant stops are unclear.

In the third and final phase, the batterer becomes contrite, kind and loving. He or she begs or pleads for forgiveness. Often, extended family members are enlisted to plead on the assailant's behalf for understanding. At least in the early years of a relationship, most assailants genuinely appear to be contrite, and honest in their beliefs that they can and will stop. Sadly, the true experience is more like that of Margarite in our chapter vignette. Most victims do not see the cycle to the violence. They see it as a random event that is over. Thus the cycle of violence continues, the ugliness of the injuries grows, and despair becomes a daily visitor and home companion.

Impact on the Victims.   The impact of physical abuse on the victim is extensive and painful. Physical injuries are serious, and may include broken bones, internal injuries, lost teeth, bruised eyes and other facial features, miscarriages, hair pulled out, and general bodily aches.

To these physical injuries must be added the psychological pain of fear and anger, as well as sadness, grief, helplessness, hopelessness, and continuous depression. Since the victim does not understand the cyclical nature of the violence, he or she remains without reasonable mastery of the situation, and has a diminished ability to find helpful solutions. Anyone who has worked in an emergency room is astounded by the capacity of spousal victims to minimize physical and psychological pain and delay seeking treatment.

Victims, and their children who witness spousal violence, are also learning terrible and inaccurate concepts about caring attachments. They are learning that those who love you are likely to hit you, and that violence is permissible when nothing else seems to work. If new and more adaptive lessons are not learned, these victims will use the only skills they know when they enter into adult relationships, and the seeds for intergenerational transfer of violence in some are sown.

The degradation and humiliation associated with physical abuse destroys self-esteem and any sense of personal significance that the victim may have had to start with. To be continuously pummeled and repeatedly injured leaves the victim feeling less valued than a stray animal. With so much energy focused literally on survival, victims of physical abuse have little time to search for meaningful alternatives.

Types of Domestic Physical Abuse

Spousal Abuse

Who would do such a thing to one's spouse? What kind of person allows such violence to occur? What happened to the marriage vows of love and commitment? What kind of violent acts are we actually speaking of? Why doesn't the victim leave for self-protection? Researchers have begun the inquiry that over time will help us to unravel the answers to these questions. Their efforts to date have provided some helpful beginnings.

When and Where?　　There are some remarkably consistent findings on these issues. Much spousal battering occurs between 8:00 PM and 11:30 PM. It occurs mainly on weekends and holidays. Most of it occurs in the kitchen, and a hair brush is the most common method of assault. The assailant is usually extremely intoxicated. Spousal murder occurs when the battering has been excessively severe, and the bedroom is usually the site of the murder. Women are murdered because there is no escape; men are murdered when they are distracted or have fallen asleep, and the battered spouse has a chance to strike back.

The Means of Violence.　　Sociologists William Stacey and Anson Shupe (1983) were studying cases of domestic violence and decided to keep a list of household weapons that had been used in various acts of domestic violence. Their partial list (Stacey and Shupe, 1983, p.30) included:

> pistols, shotguns, knives, machetes, golf clubs, baseball bats, electric drills, high-heeled shoes, sticks, frying pans, electric sanders, toasters, razors, silverware, ashtrays, drinking glasses and beer mugs, bottles, burning cigarettes, hair brushes, lighter fluid and matches, candlestick holders, scissors, screw-drivers, ax handles, sledgehammers, chairs, bedrails, telephone cords, ropes, workboots, belts, door knobs, doors, boat oars, cars and trucks, fish hooks, metal chains, clothing (used to smother and choke), hot ashes, hot water, hot food dishes, acid, bleach, roses, rocks, bricks, pool cues, box fans, books, and, as one woman described her husband's typical weapons, "anything handy."

Other social science researchers (Browne, 1983; Gillespie, 1989) have kept similar lists. They have reported that men and women also use their hands and legs to punch, slap, kick, choke, pull hair, bite, hurl bodily, and throw objects such as dishes, furniture, and scalding liquids. The violence can include stalking the victim, tying up the victim, torturing the victim, and sexually abusing the victim in the full range of ways that we noted in the previous chapter on rape. Stitches from previous assaults have been torn out; bones broken in previous violent episodes have been deliberately broken again. Some victims have even been choked or beaten into unconsciousness.

This litany of violence is truly painful to reflect on. So much sadness, so many shattered lives. And all of it kept secret from extended family, friends, and the neighbors.

For Better or For Worse. Who Does Such Things?   Most of the research to date has been done on husbands. In addition to the general characteristics noted earlier, we have learned that these men often have a history of being a direct victim of—or witness to—abuse as children, that they are often jealous and fear being abandoned, and that they use alcohol or drugs excessively. These men are often hostile, short-tempered, and easily take offense.

Many have continuing problems with reasonable mastery and attachments, and seek to overcontrol others. For example, many beat their pregnant wives when faced with sexual frustration, transitions in the nature of the relationship, and in fear and resentment of the child to come. Their skills for effective domestic problem-solving are minimal, and their use of sex and aggression are often attempts to shore up poor self-esteem, and relieve feelings of depression.

Who Are the Victims?   Most of the victims are females and, not surprisingly, they, too, often have a past history of childhood sexual or physical abuse. Like the assailants they, too, often have limited mastery skills in the face of life stress, and problems in forming caring attachments. They tend to hold traditional views of marriage and the family, and often see themselves as responsible for the batterer's actions.

These victims tend to use a good deal of denial. They deny injury, they deny being victims, they deny that personal issues interfere with the marriage. Rather, they see the marital strife as beyond the control of the couple, and attribute the spousal abuse to frustration, life stress, or excessive use of alcohol. Many do care for their abusive partners.

The constant cycle of violence will lead to certain psychological changes in the battered spouse (Walker, 1979). The female victim comes to perceive herself as helpless. Her ability to think clearly becomes inconsistent because of the inconsistency and unpredictability of the violence. It becomes hard to concentrate and to figure anything out. After a while, the victim focuses mainly on survival, loses the ability to think about other alternatives, and develops a high threshold for violence. The cycle of violence ends up seriously limiting the victim's capacity to reason clearly.

But Why Don't They Leave?   Victims remain for many reasons. They stay for the children, and do not realize that the children's witnessing of the violence is usually more damaging in the long run. They stay because they misunderstand the cycle of violence and want to believe all the statements of contrition and love in the third phase of the cycle. Others stay because they are socially isolated, have no other economic resources, or are too depressed to find a way out. Some stay because they are victims of the "Stockholm syndrome." Some stay because they love the batterer, and hope that change for the better is possible. Others stay because they realize they are getting older, and it will be harder to find another mate. Still others stay because they are so focused on survival that developing a plan to leave safely never crosses their mind.

At a more basic psychological level, still others stay because in their own minds they believe they have a way of predicting the batterer's behavior. This makes the situation more tolerable because it gives the illusion that the victim can contain the violence by acting more quietly, leaving the house when the batterer comes home drunk, and so forth. The cycle of violence, however, will continue and will escalate as we have noted. Sooner or later, the victim's theory of prediction is disproven. The assailant is too drunk one night, the violence is more extreme than usual. Something happens. The theory of prediction falls apart, and, at this point, the victim usually finds a way to leave.

Extreme Cases.   In very severe cases of abuse after years of physical injury, sexual abuse, and continuous psychological degradation, the victim may murder the spouse (Blackman, 1989). In 1986, four percent of the spousal homicides were committed by women against their husbands, with seven percent of the husbands killing their wives. This is about the same ratio as boyfriend-girlfriend homicides for the same period. Males threaten to kill if the female threatens to leave. Females kill in self-defense.

Female victims who are severely beaten for protracted periods of time and who are constantly threatened with death develop the psychology of survival coping noted earlier. There are changes that limit cognitive functioning as all energy is focused on staying alive. These victims perceive all efforts by family members, the police, and the legal system to protect them as fully exhausted. Faced with the continuing fear of death they see killing the spouse as the only solution to end the terror. In such cases, the female victim usually waits until the physically abusive spouse is asleep or not paying attention. She then shoots him from behind, puts down the weapon, cries, calls the police, waits until they arrive, and then turns herself in.

The frequency of these homicides by female battered spouses has been declining as better policing procedures and battered shelters have become available. Such homicides raise issues of imminent harm, excessive force and premeditation, and we shall return to these questions in the section on legal matters in chapter 9 as we study how society is attempting to integrate the effects of long-term battering on the human mind.

Child Abuse

Although child abuse has been noted throughout much of human history, it was not until Dr. C. Henry Kempe described the Battered Child syndrome in 1962 that any systematic study and treatment of child abuse was begun in earnest. Several authors (Justice and Justice, 1976; Kempe and Kempe, 1978) have studied these matters in detail, and here is what they have learned.

Child abuse is defined as the willful infliction of harm on one's own children or stepchildren (from a second marriage by the mother), or a live-in situation with a significant other. There are four basic categories of abuse. The first is physical abuse and includes battering, burning the child's buttocks on a gas or electric range, tying the child up, overdosing the child on medicine or drugs, and, unfortunately, many of the other forms of physical abuse that we have noted for adult victims. The second category is comprised of the various forms of sexual abuse, including incest, as we discussed in the last chapter. Physical or emotional neglect—where the child fails to develop properly because of lack of proper parental attention—forms the third category. Not feeding, bathing, or changing the child, being away from home so that the child has no companionship, emotional support, or help in daily problem-solving are common examples of neglect. The final category is emotional, hostile verbal abuse of a child. Such verbal hostility far surpasses any error the child may have made, and usually involves the complete denigration of the child rather than being confined specifically to the child's error. The child quickly learns self-hatred rather than seeing him- or herself as a worthwhile person who also happens to make mistakes. Even though there are no broken bones or bruises, emotional verbal abuse can cause equally long-lasting psychological scars.

The signs of the various types of child abuse may include cheek or body bruises, broken bones, or unusual marks on the child's body such as come from being whipped with an electrical cord, burned with lit cigarettes, or being burned by the elements on the kitchen stove. Frequent accidents can be signs of active abuse, and a distended stomach or unkempt appearance may be signs of neglect. Aggressive behavior, fully passive behavior, social isolation, the inability to make friends, and fighting with other children are also common signs of violence at home which is directed against the child, or to which the child is a witness.

Family Dynamics.   Which parents are likely to become abusive? Just as the research on incestuous families may not be truly representative, so the same may be true of our understanding of the paths to child abuse. Some medical researchers feel that some parents have not had adequate care as a child and are unable to be emotionally supportive to their own children, and therefore lack the necessary skills for child-rearing. Some health care providers see the probability for abuse increasing when poverty, social isolation, or chronic illness are present in the parents (Gelles and Straus, 1988).

Notwithstanding how it starts, here are some of the general characteristics of parent/child violence that have been identified thus far. The child is usually under age four, the abuse is repeated many times, and the period of abuse lasts from one to three years. It is the mother who is the most common abuser, and she is in her mid-twenties. The father, who may himself be violent against the mother, is about thirty years of age. As with spousal abuse, the hairbrush is the most common weapon.

Studies of parents who abuse their children have taught us something of the life stress encountered by these men and women. In two-parent families, it is common to find that both parents were direct victims of abuse themselves, or witnesses to such violence. The parents tended to marry young, had poor parenting skills, and were socially isolated from their own families. The parents' own marriage relationship had difficulties in the areas of finances, sexuality, and mutual support. Often one or both parents used drugs or alcohol as adults. Finally, one or both parents viewed the abused child as unwanted (often such children are born out of wedlock), and/or expected the child to meet the parents' needs and not vice versa. Many of the parents were depressed.

Single parents who abuse face different forms of life stress. A shortage of time with too many responsibilities leaves the single parent with role overload, and makes the parent more vulnerable to the child's behavior problems, which might include messy feedings, soiling, or intractable crying. Such single parents often work at low-paying jobs and face severe financial pressures. Additionally, they may be having problems with the child's father or their boyfriend, and have a curtailed social life outside the home. Depression is also commonly found in single parents.

These various life stresses prove overwhelming for the parents, and the child becomes the focus of the rage and frustration of daily life. The abuse begins, and the family denies or minimizes the problem, but at the same time avoids others so that the secret is hidden.

The secret can affect family life in one of at least three ways. Some families become enmeshed, that is, their lives revolve mostly around other family members. They become fully socially isolated and view the world as dangerous and to be avoided. These enmeshed families go it alone and rely only on family members. In other families, the members become disengaged from one another. At least one parent feels fully overwhelmed and emotionally disengages from family life and care of the children. The parent remains legally in the marriage, but avoids active involvement in family life. The third family system is a chaotic system where each person is on his or her own and any organized family functioning has been abandoned.

As we have noted, not all parents abuse their children when faced with stressful life events. This list for predicting possible child abuse can be narrowed to four factors. An abused parent (1) who sees the child as a disappointment (2) who has no effective lifeline for help (3) and who is in a crisis (4) may become abusive (Kempe and Kempe, 1978, p.24).

These risk factors are not presented to frighten parents, but to alert all parents who feel overwhelmed by their children that these are the circumstances in which there is an increased likelihood of striking. It is equally helpful to remember that these factors can each be addressed so that the probability of abuse can be markedly decreased. Many parents, for example, have found it helpful to call a local parental abuse hotline or to attend self-help groups for parental abusers. These strategies soon lead to reductions in abuse and general family confusion.

Impact on the Victims. Untreated child abuse may lead to painful, chronic, lasting impairment to the child's physical and mental health. The more the abuse, the greater may be the subsequent impairment. A recent study of children who witnessed, but were not direct victims of spousal abuse found that witnessing verbal abuse led to mild conduct problems, witnessing verbal and physical abuse led to more severe conduct and emotional problems, and witnessing verbal abuse with physical abuse that necessitated going to a shelter had the greatest negative impact on these children. It led to higher levels of emotional problems and lower levels of social interaction. We can assume being a direct victim would lead to similar painful outcomes. Children do not grow out of these severe disruptions in normal development. The consequences last a lifetime, and both the child and society lose.

From a medical viewpoint the child can end up with serious bone, joint, and muscle problems as well as lasting consequences from internal injuries and surface bruises and cuts. There is increasing evidence of possible damage to the child's central nervous system and brain functioning, even when no head injuries were sustained in the abuse. Some end up with mental retardation, and others have problems with paying attention and with abstract reasoning. Many use language less frequently. Beating a little child may well alter the child's biology for life (Siegel, 1999).

Children fare poorly in the psychological domain as well. Understandably, many of these youngsters have the symptoms and feelings associated with PTSD. Anxiety, hypervigilance, episodic panic, nightmares, intrusive memories including dissociative materials, helplessness, withdrawal from others, and depression may all be present.

Skills for developing reasonable mastery and caring attachments are often inadequate because the adult role models at home are poor, and the learning of these fundamental building blocks is never mastered. As youngsters, they have problems making and keeping friends, and often have academic problems at school. As teenagers and young adults, they may fail at their career potential, become involved with substance abuse and promiscuity, or become further withdrawn from community life. A few will become involved in crime, some ten to thirty percent will abuse their own children, and many will have lifelong problems with authority figures.

These victims will be followed by continuing fears of being humiliated and abandoned. Many will have a basic and pervasive mistrust of the environment and will always be concerned with issues of physical safety. Some will seek the attainment of money as a goal in life to ensure the safety and support that they are unable to find in caring relationships. Many will be afraid of becoming violently aggressive themselves. Many will remain unassertive and not address their own emotional needs. Others will remain phobic of all forms of violence, and avoid television programming, sporting events, and the like. Many will be haunted by guilt, self-blame, shame, and the fear of being re-victimized. All of which takes its toll in poor self-esteem and which may lead to being chronically depressed.

* * *

This chapter chronicles an equally dark night for the human soul. Each of us wants to be loved, yet here again we see how traumatic violence can damage the universal dream of being loved and accepted for who we are. In physically abusive situations, marriage vows ring hollow, one's own parents become the enemy, and society offers little in the way of help because the privacy of family life is sacrosanct. It is frightening. It is confusing. It is depressing, but as we shall see, it need not be one's painful legacy for life. As in cases of sexual abuse, victims of physical abuse can change. Victims do grow. If you have been battered, there is hope. It need not be this way.

7

Combat/Family Alcoholism

Man has discovered death.
— Yeats

These pitiful days will come to an end.
— Anne Frank

Perpetual motion: it was the family joke, but it was also true. Henry could not sit still. He poured his considerable energies into his studies, into varsity sports, into visiting his friends' families. Anything to escape the tension at home. He also had to referee his parents' frequent arguments, raise his siblings, and pacify the neighbors. If he tried to rest, he became depressed. It was easier to remain on the move.

He wanted out, and the Marines wanted a few good soldiers. His father, who had never really been of any help, had been vehemently against his enlistment. But now Robert stood tall and proud, and swore his oath. His mother fought back her tears. His father never showed up. Typical.

"You do what I tell you or I'll beat the hell out of you. This is my house and don't you ever forget it. If you don't like it, go live on the streets." Henry's father was drunk again.

Henry remembered years of arguments, years of being embarrassed, years of being tense, and still more years of verbal abuse and scapegoating. As a little kid he used to force himself to sit completely still and immobilized so that no body movement of his would serve as an excuse for his father to drink. Henry had tried to hide the alcohol, to pour it down the drain. He had tried anger, reasoning and prayer. But God did not answer. Nothing worked. His father's drinking became progressively worse.

His mother had solved the problem by going on strike. She returned to her job as a nurse and did little at home but cook. His younger brother was doing drugs, and his sister was dating someone as old as their father, for God's sake. Henry did what he could, but he sensed he was failing. All the neighbors knew. Why didn't they help? Why didn't his mother just leave the old drunk?

He would leave for bootcamp in the morning, but first he was determined to find out why his father drank. Henry found him in the kitchen where he was nursing a can of beer. Henry was hoping to keep the lid on his bitterness and his sense of rejection as he asked the question that had haunted him for all of his teenage years.

There was a pause. A deafening crash of silence, actually. His father did not erupt in rage as he had expected. He wasn't even really angry.

"I started drinking after the war. I saw my best friend killed in combat in Vietnam. He was blown apart right in front of me. One arm. One half of a leg. That was all that was left. I gave the body parts to the medics to return to his family. I've never been able to get over it. Alcohol helps me forget."

The lump in Henry's throat thickened. Both men sat in silent tears in the darkness of the night.

*   *   *

Henry's family illustrates the two types of traumatic events to be studied in this chapter: combat and family alcoholism.

Unlike sexual abuse and battering, there is somewhat less of a societal stigma associated with having combat-related PTSD, and there is a growing national awareness of the deadening impact of family alcoholism on family members. In these two cases the country has a better understanding that veterans and the children of alcoholics did not personally elect to become victims.

The impetus for such increasing awareness began in the 1970s from the combined efforts of the women's movement to understand violence, and the burgeoning self-help recovery movement, particularly the National Association for the Adult Children of Alcoholics. All of these groups have focused on the nature of these traumatic events, and on the negative consequences that may ensue for untreated victims. With less stigma has come more accurate recordings of such events, and less shame in describing what has actually happened. Hopefully, the stigma associated with sexual and physical abuse will likewise continue to lessen, so that more can be learned about how to similarly assist recovery in those painful circumstances.

We shall begin this chapter with an examination of combat and war and its impact on human functioning and health. We shall briefly review the normal tasks of adolescence, and see how war experiences may significantly alter the normal course of development. We shall also focus on prisoners of war, both military combatants as well as civilian noncombatants. The chapter concludes with the effects of parental alcoholism on children as they mature into adulthood. We shall look at the family dynamics, the unspoken rules, and the range of mixed messages that may negatively affect these offspring in profound ways.

As with coming home at the end of a war, physically moving out of your alcoholic parent's home does not necessarily resolve the negative effects of the painful experience that the victim has endured.

Combat

There is another battle that often occurs after the war is over and the cease-fire has been signed. It occurs in the mind of the warriors, and we know it as posttraumatic stress disorder.

The emergence of PTSD as a serious clinical problem affecting many soldiers and requiring special steps for recovery came in part from studies of veterans of earlier wars. Sometimes referring to it as "shell shock" or "traumatic war memories," many medical observers over the decades have described a chronic pattern of behavior that we now refer to as PTSD. In several studies, investigators reporting the effects of combat on soldiers in World War I, World War II, and the Korean War noted the presence of recurring intrusive memories, nightmares, irritability, exaggerated startle response, and depression. The occurrence of such problems was more likely to be found in those military men and women whose childhoods were marked by parental alcoholism and parental discord. Before 1980, there were few studies of soldiers who were in active combat and also taken prisoner of war, but those studies that do exist suggest that similar disruptions in physical and psychological functioning were experienced by POWs and lasted for many years.

The Vietnam War caused much soul-searching in the United States both during the war and after its conclusion. It led to debates about the nature of war and what should be the extent of the country's involvement around the world. It led to changes within the military that resulted in high-technology weapons, an all-volunteer armed service, and much greater formal training. Equally important, it also led to a study of the mental health of those who fought in the war.

In the mid-1980s the United States government decided to conduct a very thorough study of the effects of war on the combatants. This study focused on Vietnam-era veterans, and clearly demonstrated that for some combatants war is bad for mental health.

Although a majority of the Vietnam veterans made a successful reentry to civilian life after the war, the National Vietnam Veterans Readjustment Study (Kukla et al., 1990) found that a little over fifteen percent of the men and a little over eight percent of the women who served in that theater of operations were currently experiencing PTSD symptoms. The functioning of these men and women was impaired in other ways as well. The veterans with PTSD had subsequent higher unemployment, more divorce, more civilian violence, more homelessness, more substance use, more depression, and more chronic illnesses than would be found in a comparable civilian population. More than twenty years later, the negative effects of traumatic war stress remained in an appreciable number of veterans.

Research studies on prisoners of war conducted since 1980 have shown that some prisoners of war from World War II and the Korean War have the same negative PTSD effects as the Vietnam-era veterans. These negative effects of untreated trauma have persisted in many veterans for as long as forty or fifty years.

Similar studies of combat veterans in Israel and Canada also reveal the presence of PTSD in many soldiers. From the studies currently available, we find that in any army, somewhere between ten to thirty percent of the combatants in a war may develop PTSD as a result of their wartime participation. War is hell not only on the battlefield, but also later on in one's own mind.

The Making of A Soldier

The Just War. War is never an easy undertaking for the combatants or for the society as a whole. If the war is perceived as being waged for an honorable cause, however, this sense of meaningful purpose may mitigate in some soldiers the otherwise seeming futility of the severe life stress they must face. Thoughtful men and women both in and out of the military have given lengthy deliberation to the circumstances under which armed military conflict is considered a just undertaking for a society. There are pacifists who believe no war is ever justified and hawks who equally believe most wars are necessary. Many fall somewhere in the middle as society continues to struggle with this complex phenomenon. The decision to go to war is an intricate moral dilemma.

For a war to be just, several conditions are thought to be required. The war must be one of self-defense based on high moral principle where the nation's interests are at stake. It is a war to be fought only after all other reasonable economic and diplomatic efforts have been exhausted, and then only with the minimal force necessary to resolve the specific conflict. When utilizing this avenue of last resort, prisoners of war are to be housed, fed, and cared for humanely, and unarmed civilians on both sides must be protected from direct combat exposure. These are complex criteria to meet and even more difficult to implement in actual warfare, but they are important, if a society is to ask its youth to place themselves in harm's way.

Normal Adolescence and Young Adulthood. With the exception of career officers, many of the combatants in an army are late adolescents and young adults in their early twenties. To understand the impact of traumatic war stress on these young minds, it is helpful to understand the normal developmental tasks of this age group.

Psychologist Erik Erikson refers to this period of human growth as the Identity Crisis (Erikson, 1963). During these years the young adult must make some initial career choice, establish a sense of intimacy with other adults, and thoughtfully develop his or her own basic value system.

The first task for these young adults is to make some initial entry into the labor force, and to refine over time their personal career choice. There must be a balance between the individual's needs, skills, and interests and the needs of society as a whole. Finding this balance enables the individual to continue his or her personal growth, to gain some sense of reasonable mastery, and to contribute to the common good.

Each young adult must also establish some basis for caring attachments. Intimacy refers not only to sexuality but to other important aspects of relationships such as relying on others, sharing with others, being helpful to others. It is during this period that many may select the mate they choose to share their lives with. Not only is this an important choice in its own right, but it can also become a source of caring for others in terms of children and the larger community.

Finally, young adults need to develop their own ethical guidelines by assimilating what they have learned at home, in school, and in church, synagogue, or mosque. In part, young adults also incorporate the values learned from personal life events and from interaction with teachers, mentors, and the community at large. These assimilated values provide an initial meaningful purpose to life and some sense of personal significance. Most young adults are quite idealistic. They believe in God and country. They believe in marriage and family. They believe in a good and decent life. They believe in treating others with fairness and respect.

The Nature of Military Life.   What does the young adult with these normal developmental tasks encounter in the armed services?

First, the exercise of reasonable mastery is somewhat modified as the army essentially demands the soldier surrender his or her autonomy. One does what one is ordered to do. Discipline and obedience to authority, even in the face of one's own death, are the military values that one accepts because they are necessary for the army to function. One's career choice temporarily becomes the destruction of the life and property of the enemy in the service of a just cause, as this is the career choice that society expects of our countrymen for the time they are on active duty.

The normal process of caring attachments and intimacy is also altered. The soldier is physically separated from family and friends, but must protect them and his or her military colleagues so that in the longer term a decent way of life is preserved for all. But the means to attain the goal is to kill others. The young adult is asked to protect and nurture caring relationships to others by killing humans who presumably have their own networks of caring attachments. To complicate matters further, in combat training it is the soldier versus the enemy, but in the actual theater of war it is not always clear who is the enemy and who is an innocent civilian noncombatant. Since soldiers bond with their buddies for survival, they often become less involved with civilians.

Military life, by necessity, also challenges the newly formed value structure in the new recruit. One can serve God and country in military service. One can fight for one's family and one's countrymen. But is war fair? Is killing others in combat a reflection of a good and decent life? Of treating others with respect? The young soldier must struggle with these divergent aspects of life. A just war makes the process somewhat easier. The nature of the military experience challenges the normal developmental processes in many soldiers long before the first shot is fired.

The Nature of War.   War is psychological trauma. It is a potentially life-threatening event over which an individual soldier has only very limited control for personal survival. War is outside normal human experience. It is cold-blooded destruction and the killing of human life in which the humanity of the enemy must be denied. It is the active and passive witnessing of death of the enemy and of one's comrades. It is atrocities of every kind imaginable. It is body parts and blood everywhere. It is personal injury and pain. It is the continuous fear of one's own death. It is a continuous period of high physical and emotional arousal. It is a period in which testing one's self can be exhilarating as well as terrifying. War is also filth, material deprivation, and ubiquitous disease, pain, suffering, and death. It may also be drugs, alcohol, and prostitution, as combatants seek to numb the horror.

In actual warfare the young adult's concepts of reasonable mastery, caring attachments, and basic values for a meaningful sense to life are shattered. It should not surprise us that some combatants later develop PTSD.

Prisoners of War. Prisoners taken by the enemy in war are exposed to a second possible source of traumatic violence—torture.

Torture is the deliberate infliction of physical or mental pain on the prisoner by his or her captor to break the will of the captive. The intended harm with its pain and suffering are part of a process designed to extract information from the prisoner, to see if the prisoner, in the face of severe abuse, changes what he or she has already told the captor, and/or to serve as an example to others to frighten them into complying with some pattern of behavior the captor wishes.

Torture is cruel and degrading by intent, and its methods are varied and ugly. Among some of the more common practices are being physically punched, kicked, or beaten. Victims may receive electric shock to body parts including the ears and genitals. Forced standing, bondage, blindfolds, and handcuffs may be employed for long hours. Hair pulling, head banging, burns by chemicals or boiling water may be alternated with long periods of deprivation of food, water, and sleep. Sexual abuse, including rape or threat of rape, may occur. Victims may be forced to watch others being tortured or killed, and may be repeatedly subjected to their own mock execution. The guards themselves often further extort or exploit the victims.

What I have listed above is illegal by international accord, but war is war in many ways, and torture all too frequently is employed to destroy the spirit of the victim so that he or she will do as they are bid. All forms of torture are potentially life-threatening, are beyond the control of the prisoner, and may result in PTSD.

Civilian Noncombatants. The fury unleashed by war very often directly affects the lives of unarmed innocent civilians. They too are exposed to the traumatic violence of war and the possible forms of torture noted above, as their homes and lives are often in the path of destruction. Their loved ones are conscripted, taken political prisoner, or disappear in other ways, never to be seen or heard from again. Physical pain, mental anguish, and death are common in the lives of many civilian noncombatants.

While our knowledge is far from complete, some scholars have been investigating the impact of war on such victims. Psychiatrist Robert Lifton (1967) has studied the Hiroshima survivors, and many authors such as psychiatrist Joel Dimsdale and his colleagues (1980) have examined the painful consequences of the Jewish holocaust in Germany.

Dr. Lifton has catalogued at length the aura of death that enveloped the victims after the first use of atomic weapons in war. Death has been the constant companion of these victims for over forty years. First came the immediate deaths of their loved ones and neighbors and of a way of life that had been theirs. This was followed by the death of the earth through contamination. Flowers, trees, animals died. Water was poisoned. The soil was contaminated and crops could not grow.

Death continued to follow these victims over time as the effects of radiation led to severe and continuing medical problems and birth defects in subsequent generations. The impact of such extreme death was so pervasive that Dr. Lifton (1967) found that the victims sustained a lifelong identification with the dead (a form of severe survivor guilt), to maintain links with the deceased and to honor their memory.

The survivors of the concentration camps in Germany fared just as poorly. Those who survived did so by using denial, detachment, rage, informal support groups within the camps, hope, and belief in God. Their survivorship is extraordinary, but it did not free them from the long-term consequences of abuse (Dimsdale, 1980). Many of these victims developed what is referred to as the *survivor's syndrome*. It includes anxiety, the loss of initiative and a phobia of violence as well as social isolation, problems with sexuality and parenting, depression, and total lack of joy. The painful memories of the past are always nearby in memory.

Medical researchers have begun to document the appearance of the survivor's syndrome in the children of these victims. Children who were not even alive at the time the parent was being abused commonly have hostility, depression, guilt, separation anxiety, and intense Jewish identification. Many are depressed.

The mechanisms of this intergenerational transfer of the negative consequences of parental traumatic abuse are clear. Perhaps these children of survivors identify with the parents' suffering, possibly they model the parents' coping behavior. The answer awaits further research, but what the Japanese from Hiroshima and the Jewish victims in Germany teach us is that the effects of traumatic abuse can be equally as devastating and long-lasting for the civilian noncombatants as they can be for the actual warriors.

The Impact of War on Those Who Fought. In addition to physical injuries and loss of limbs, the classic symptoms of PTSD may be found in many returning veterans. The symptoms of physical arousal, intrusive memories, and avoidance complicate their civilian work and home lives. Divorce and underemployment are common, as we noted earlier.

Many feel guilty and engage in acts of destructive behavior against the self and others. Many self-medicate the continuing distress of war with alcohol and drugs. Many have psychosomatic illnesses and more chronic health problems as they age. Many remain depressed, and a sense of well-being is chronically absent.

Prisoners of war have been found to have the same debilitating effects in their own subsequent civilian lives. There is increasing evidence that untreated POWs have more anxiety, more depression, more alcohol abuse, and more hospitalizations than those soldiers who were not captured. We do not have enough research, as yet, to fully ascertain whether combat plus torture results in greater impairment than either combat or torture singly.

It is remarkable that as many soldiers as do return to a normal civilian life. Such successful reentry is a testimony to the human spirit. Perhaps these men and women utilize the skills of stress-resistant persons, and have somehow learned to exercise reasonable mastery, caring attachments, and a meaningful sense to life in the midst of human hell.

It is not surprising either that many warriors do develop PTSD. War is a major assault on the normal developmental tasks of youth, and the firm grounding for health and well-being in adult life is never completed in some. To face these limitations and correct them is to survive. And to survive, for any victim, is an achievement of note.

Family Alcoholism

Family alcoholism is best understood as a pattern of thinking, feeling, and behavior that emerges as a way to cope with life stress and the daily events in a home with active parental drinking. The parent's alcoholism, which has gotten him or her into trouble, affects the spouse, children, and any other permanent resident in the home, such as a grandparent. Parental drinking creates a distorted way of viewing the world and faulty ways of coping with everyday problems, such as denying or minimizing serious family conflicts. If left untreated, the poor coping skills and faulty assumptions about life learned in the alcoholic home when one is a child will follow the adult for all of his or her life. Physically moving away from home will not change it, and the mere passage of time will not correct it either. Untreated children in alcoholic homes become Adult Children of Alcoholics (ACoAs) until the system of beliefs and methods of communicating and problem-solving that they learned in the alcoholic home are corrected toward more normal ends.

There are actually four types of alcoholic families. The first is the most obvious. This is a family with an active parental drinker. Family life revolves around this drinking behavior in ways that we shall examine shortly. The second type of family is one in which the alcoholic is no longer drinking, and has become physiologically sober. Without some form of treatment, however, the family's maladaptive ways of coping, learned during the period of active drinking, will not be corrected. This same process is true of the third family type. In this family constellation, the grandparents were the alcoholics. Their children choose to remain abstinent to spare another generation of the family the chaos they had to live through. Again, if there was no treatment beyond self-imposed sobriety, the alcoholic family thinking will not likely be corrected, and their own children, the grandchildren, may become alcoholic. The final type of alcoholic family results in situations in which a healthy nonalcoholic family has a child that later drinks alcoholically. The child's behavior will impact on the family to produce the same confusion as the other types of family alcoholism.

The faulty approach to life learned in any of these family types will never be corrected in and of itself. It is a psychological way of responding to the events of daily life, and Adult Children of Alcoholics will carry it in their minds wherever they go until conscious efforts are made to correct it.

The Extent of the Problem.   The negative health effects of parental alcoholism are again epidemic in number. One in eight

Americans is an ACoA. This is almost thirty million men and women. As many as half of these ACoAs are likely to become alcoholics themselves, and many others may go on to marry alcoholics and enter into the family chaos they tried so hard to remove themselves from. ACoAs are more likely than their friends from non-drinking families to drop out of school, more likely to abuse other drugs, more likely to have eating disorders, and are more likely to attempt or actually commit suicide. Many remain depressed for years. (See Woititz, 1983.)

The unpredictable behavior of the intoxicated parent produces in the child a state of continuous emergency physical and psychological arousal. In addition, over half of all family violence occurs in alcoholic homes, and this violence includes physical and sexual abuse as well as emotional neglect.

Not every child who grows up in an alcoholic home will become an adult child in mind and outlook. Many do, however, and the impact of the problem on their individual lives and on the welfare of society as a whole is enormous.

Family Dynamics: The Alcoholic Home

A home with family alcoholism is marked by continuous unexpected behavior and general confusion. This is due fundamentally to the parental ingestion of alcohol which alters the parent's normal brain chemistry. The adult's motor behavior is changed, and may result in poorly coordinated walking, slurred speech, and the inability to chew food smoothly among other things. The alcohol affects the limbic system and the temporal lobe where memory processes occur. This can lead to sudden and unexpected feelings, such as hostility and anger, as well as trouble remembering the next morning what happened the evening before. (This memory amnesia is known as a blackout.) Finally, continued use of alcohol changes the addicted person's thinking and values. Reasoning becomes inaccurate and alcoholic individuals begin to treat others with meanness, such as deliberately embarrassing them, stealing from them, and generally neglecting them. Many are accomplished manipulators. Twenty years of heavy drinking has also been known to lead to some forms of subtle harmful neural changes in the brain, although some of these neural changes appear to reverse themselves after eighteen months of sobriety.

Additional family confusion arises as the family members attempt to cope and respond to these forms of problematic behavior. Most of the family's efforts are directed toward containing and minimizing the impact of this disruptive behavior. Many alcoholic homes are marked by some basic rules for survival, and a series of mixed messages when the rules fail as they usually do. It is a painful way for a child to live.

The Rules.   There are four unwritten rules that usually govern the interaction and functioning of members of an alcoholic household (Kritsberg, 1988). The first rule is one of rigidity. Since alcoholic behavior by its chemical nature is unpredictable and out of control, family members respond by seeking to exercise some form of mastery. Attempts at discipline, homework, household chores are often imposed in an effort to gain control over some aspect of family life in the hope that this may contain the alcoholic's unpredictable behavior. Such attempts are futile because the chemically altered brain of the alcoholic is by definition not in control of its normal faculties. This continuing cycle of attempts to contain what is truly not containable results in ever-increasing rigidity in daily routines. Often it happens that suggestions from others that might lead to helpful modifications of the family's way of coping are summarily dismissed solely because the suggestions are new and outside the usual rigid routines.

Secrecy is the second unwritten rule. Silence about the parent's drinking to anyone outside the home is insisted upon. ACoAs often develop intense shame about the parent's drinking because of this stance of silence. The silence is kept so as not to embarrass the family; but this silence leads to isolation from others, and new ways to cope with life stress, including family alcoholism, are never learned. Personal growth and the capacity to respond flexibly to change are stunted.

The third rule is denial. Alcoholics and their families deny that there is anything wrong. They all walk around this huge elephant (alcoholism) in the living room. They deny the problem, deny its impact, and deny that they are not a normal nonalcoholic family. Such processes of denial carry further important consequences because ACoAs begin to deny their own feelings of hurt, anger, sadness, and grief. As they become older, they are unable to identify accurately how they feel. As a result, as adults they often don't know what their personal needs are.

The fourth rule is one of isolation. To keep the family secret and to keep from being embarrassed by the alcoholic's behavior, ACoAs socially withdraw and isolate themselves from others. Such behavior has many later interpersonal complications. ACoAs have problems making friends because they have not learned the skills to do so, and they have problems in intimate relationships because the expectation is to remain isolated. Often they never learn to have spontaneous fun and to be relaxed. One must be alone and stay in control.

Mixed Messages.   Because of these four rules and the general inconsistency in alcoholic behavior, children in these homes encounter mixed messages from the parents. These messages are confusing, contradictory, and beyond the capacity of the child to resolve. They may leave the child anxious and overwhelmed. Here are some of the common ones.

The child will be told to always tell the truth. Yet one or both parents will continually lie and distort because of the attempts to use denial. The child will be told that he or she is loved, but promises to spend time with the child will be broken, and the child is then left in a double bind. Birthdays will not be observed and so forth, as the parents are not consistently present to nurture the child. The child will be told that he or she is important, but often the child is actually subjected to physical or sexual abuse and emotional neglect. Children of alcoholics are repeatedly exposed to harsh and severely critical verbal abuse. They will be told they can do nothing correctly, that they are not lovable, and that they have no future to speak of. These criticisms are really the beliefs the alcoholics have of themselves, but displace onto their children. The children are in effect being brainwashed, as they are receiving only the messages of the alcoholic parent. In time the children come to believe these things to be true of themselves. Paradoxically, these are often the same alcoholic parents who cling dependently to these same children to resolve family matters beyond the alcoholics' capabilities to address.

Children in these homes respond to these four rules and the system of mixed messages in differing ways. Some become super-responsible, super-controlling, and overachievers. Some try to keep the family peace by placating others and ignoring their own needs. Some remain isolated and lonely. Others become angry and get into trouble outside of the home to punish themselves and their parents.

Often they become true codependents. A codependent is one whose behavior and attitudes are continuously determined by another person. Codependents become preoccupied with meeting the needs of the other and building their lives around the other because they are extremely dependent on that person. Codependents have little healthy sense of self. Often they may be seeking financial security, nurturance and caring, or enhanced self-esteem. Codependents are not honest about what they see or hear or feel. They distort reality so that reality will meet the needs of the other. Such distortions further distance the codependent from his or her own true feelings and needs. If this continues for a long enough time period, the codependent will truly lose the sense of self, and may have problems finding a way out of the subsequent confusion. Codependency is not a satisfactory substitute for independent living.

Why does this happen? How does the process of codependency begin in a family with a victim member? If we reflect for a moment on what we know about codependency, we may see the beginning of an answer. You will remember that codependents are anxious, fear being abandoned, believe they are unlovable, and have poor self-esteem. These are some of the characteristics that we have noted to be true of victims. In truth, many codependents are victims of untreated trauma themselves, and they have an uncanny ability to date and mate other victims. Adult Children of Alcoholics marry alcoholics. Sexually addicted persons marry into yet another unhappy family situation. Battered children grow up to marry physically abusive spouses.

Why marry into it again? Wouldn't someone realize from the first time how miserable life was?

This seeming puzzle may not be so hard to understand. Let us use family alcoholism as an example. When you grow up in a dysfunctional home, you learn certain rules of family life, as we have seen. Don't rock the boat, lie low lest you be hurt, don't have feelings, keep the family secret, don't trust others, and so forth. One learns to speak a specific family language that is quite different from the language of normal families, which are more open and caring. When children then grow up and begin to date in search of their own future mates, children from dysfunctional families still speak a language that is different from that of children from normal families. Since there is little meaningful communication with peers from more normal families, teens from dysfunctional families keep dating till they find someone whom they can understand and feel comfortable with—someone who speaks the same language. The familiar language, of course, is that spoken by a teenager or young adult from another dysfunctional family. Both young adults know the "rules" of dysfunctional family life. Early on they feel quite comfortable with one another. The tragedy, of course, is that these two young people with their dreams and aspirations for a better, more loving future do not have the skills to cope adaptively with independent adult life, marriage, and child-rearing. In short order, their hopes fail, and their dreams become ashes.

For all of these reasons, children in alcoholic homes learn at a very early age not to think, not to feel, not to talk, and not to rely on others. Such strategies do not lend themselves to the development of reasonable mastery, caring attachments, and a meaningful purpose in life. Thus as adults, ACoAs spend a good deal of time trying to ascertain what normal behavior is, constantly seek approval, judge themselves mercilessly; and rarely enjoy adult life.

Health Consequences.   Having an alcoholic parent places the offspring at increased risk of a full range of physical and psychological health problems, both in childhood and adult life.

They may have any of the medical consequences associated with physical or sexual abuse if such abuse has occurred. In addition, ACoAs may have problems with psychosomatic diseases as a result of the sustained emergency mobilizations that home life necessitated. Anxiety states including agoraphobia, panic, and obsessional worry may be present. Addictive behavior is common, including problems with alcohol, drugs, eating disorders, and sexual addictions. Clinical depression and thoughts of suicide may also be present in many cases.

Adult children often have continuous distress at work and in their personal and sexual relationships. Feelings of anger, guilt, shame, embarrassment, loneliness, and sadness are constant com-

companions. ACoAs usually do not know how to play and have fun, and happiness eludes them.

Will it always remain this bleak if you are an adult child? It need not. Many ACoAs have successfully overcome their childhood confusion to lead successful and happy adult lives. Some have used self-help groups, others have sought individual counseling and still others have studied more normal, adaptive families as strategies for overcoming the legacy of family alcoholism.

Posttraumatic Stress Disorder and Family Alcoholism

Does parental drinking cause psychological trauma and the symptoms of PTSD? In the absence of needed research, the medical community remains undecided on this issue.

Unless an ACoA has been sexually or physically abused by an alcoholic parent, most ACoAs do not have flashbacks of parental drinking. Some researchers see these flashbacks as the key marker in psychological trauma. Although they would agree that family alcoholism is a miserable experience, these researchers would say that parental alcoholism is not a true traumatic event because ACoAs do not have family drinking flashbacks.

Other researchers would be quick to point out that ACoAs often have all the symptoms of psychological trauma except flashbacks. They would point to the hypervigilance, exaggerated startle response, continuing sleep disturbances, worry, intrusive memories, avoidance of others, and the presence of a wide variety of addictive behaviors as evidence that ACoAs are indeed victims of trauma.

Based on the available medical evidence, none of us is presently able to answer this question fully. In over thirty-five years of counseling my experience with ACoAs suggests that many in fact appear to be victims of untreated PTSD as a consequence of parental alcoholism. While it is true that they do not have true flashbacks, it is also true that they do have every other symptom of PTSD, including a morbid intrusive preoccupation with the family drinking problems. It is likely that over time the diagnosis of PTSD will be broken down into different types of subgroupings, and that the effects of family alcoholism on its offspring will be in one of those groupings.

* * *

This completes our review of the major syndromes of traumatic events that have been formally researched. The review has not included sudden major illnesses such as heart attacks, car accidents, street crime, natural disasters, and sudden unexpected death of loved ones because less is known in these areas. To be sure, these are traumatic events for many, and undoubtedly we will learn more as community and professional interest in PTSD increases.

This review presents further evidence of the presence of addictive behavior as the result of traumatic events. *In each of these four types of trauma the research literature continuously notes the present of addictive use as a long-term consequence of untreated trauma.* Other medical diseases like heart disease, cancer, or diabetes do not lead to such addictive behaviors, but psychological trauma and PTSD do. Findings such as these again point to the importance of treating both the addictive behavior and the traumatic event in victims who suffer with both problems.

After reading this catalogue of misery, some of you may be feeling discouraged, but there is no need for such pessimism. Many victims have recovered from posttraumatic stress disorder by implementing the adaptive skills of stress-resistant persons in their own lives, and you can do so also. We now turn our attention to the helpfulness of stress-resistance as we outline the steps to recovery in the final section of the book.

Part 3

POSTTRAUMATIC STRESS DISORDER: STEPS TO RECOVERY

8

Recovery: Help for Victims

Hello darkness, my old friend.
— Paul Simon

Remember in the depth and even agony of
despondency that you will feel well again.
— Abraham Lincoln

Mrs. Ellis walked slowly from the Senior Citizen Center luncheon toward her apartment. The air was clear and crisp. The mountains in the distance retained their snowy crowns as the aspens turned their golden hue of yellow. Her Benjamin had brought her to these mountains several decades before and she had grown to love them both.

Benjamin's illness in the forty-ninth year of their marriage had been particularly difficult for her. In spite of her own poor health, she had kept him at home during those years of his failing health so that he could die in her arms. The instant of his death had shattered her soul for almost two years and her widowhood had been yet another study in how loneliness ages the human heart.

Their life had been difficult. Money had always been scarce in those early years. To carry to term their only child she had had to lie flat in bed for the last seven months of the pregnancy. But love could make it easy, and they had striven for perfect love. Gradually they put the pieces together, raised their daughter Carolyn whom they deeply loved, and helped others in need as best they could.

141

She reached for the key and unlocked the door to her apartment. Suddenly she felt a sharp sting on her left cheekbone. A second blow to her neck followed. She fell forward on the scatter rug, hit her skull on the corner of the table when she lost her balance, and now lay bleeding on the floor. Her hearing aid had been jarred loose.

The assailant came at her again from behind. Verbal epithets of loathing and disgust were followed by several blows to her lower spine. The assailant kicked her over onto her back, and Mrs. Ellis saw such hatred in the perpetrator's eyes. Repeated kicks left Mrs. Ellis unconscious.

The police questioned her in the emergency room. Did she know her assailant? No. My eyes were bloodied. Could she give a description? No. I was attacked from behind. The police left. She began to cry quietly. Mrs. Ellis did know her assailant. It was her daughter Carolyn.

*   *   *

Mrs. Ellis's experiences illustrate an increasing form of violence in our culture, that of elder abuse by one's own children. General frustration at one's lot in life, the search for an early inheritance to sustain the good life, and anger at the responsibility for an older adult in failing health are some of the reasons adult children abuse their parents. None of it is acceptable behavior, and in each case the parental victim may well develop posttraumatic stress disorder. We cannot assume that elderly persons are better able to cope with traumatic events because they have experienced other painful events over a period of many years. The sense of being overwhelmed and out of control, the sense of betrayal by a loved one, the sense of the utter futility of such violence are as intense for older persons as they would be for any of us at any age.

But life continues for all victims, including the elderly, and Mrs. Ellis must now ask herself the basic questions that each victim confronts: What happens to me next? How do I go on? How do I put these painful memories behind me? What must I do to feel better? These are important questions for those of you who are victims, and we turn our attention now to the goals and treatment steps on the path to recovery.

Recovering from Posttraumatic Stress Disorder

The Goals

As a victim, you have been dealt with unfairly by life's fortunes. You deserve, and you will want, to utilize the best strategies for coping with your adversity that you can find. A basic theme in this book has been an emphasis on the importance of the skills of stress-resistant persons (Flannery, 2003). These men and women use their effective problem-solving skills to maintain good physical health and a sense of well-being. As we noted earlier, their emphasis on reasonable mastery, caring attachments, and a meaningful purpose in life helps them buffer the stressful events of life, including the severe life stress of traumatic events.

Each of you who is a victim of violence will want to develop the skills of stress-resistant persons so that you may use them to hasten your recovery from PTSD, and then have them at the ready as you face other more common stressful situations. Both the goals and treatment steps for recovery are based on the importance of developing stress-resistance.

There are four goals for recovery that we need to be concerned with.

1) Reduced Physiological Arousal. The first goal is to restore your body chemistry to as near normal a state as it was before the traumatic occurrence. As time permits, you want to reduce any undue emergency mobilization reactions and avoid any unnecessary kindling phenomena that can lead to hypervigilance, sleep disturbance, and the like. You do not want recurring intrusive episodes nor the depression that may result in the chronic PTSD. Finally, you do not want the continuous endorphin opiate-like withdrawal that can lead to some addictions. For all of these reasons, reducing the arousal states associated with violence is the first goal of recovery.

2) Reasonable Mastery. The second goal should be the presence of some sense of reasonable mastery. You want to be able to exert some personal control over the daily events in your life, and to be able to press on in life without being interrupted by intrusive memories or having to avoid people, places, and things to keep from feeling distressed and uncomfortable. No victim wants to feel helpless in the face of the various stressful life events that happen to any of us, and you need not accept this condition either.

3) *Caring Attachments.* You want to restore or make new caring attachments to others. We have noted earlier the health benefits to be attained by such support networks, especially the natural enhancement of endorphins that might be of particular importance to some victims as one way of regulating the opiate-like withdrawal problem associated with kindling. Caring attachments to others also provide any of us with the support and information that can be so helpful in a crisis. Being with others is far preferable for victims than being withdrawn and isolated.

4) *Meaningful Purpose to Life.* Your last basic goal is to be able to make sense out of what has happened to you in such a way that it allows you to go on. Each of you needs to accept the fact that the event has happened, to grieve its loss, to examine how it has reshaped your life, and to find a renewed sense of purpose in life. These steps will enable you to review your fundamental beliefs, to restore order in the face of harmful events, and to ensure a helpful commitment to the future.

Based on these four goals I have outlined in clear, direct steps the various coping strategies that have proven to be effective for many victims.

The first section on treatment outlines the effective steps for coping with an acute traumatic episode. These steps tell you what to do if your traumatic event happened recently, and you are still feeling overwhelmed or angered by what has happened. These coping strategies are helpful for acute PTSD.

The second section deals with the recovery needs of those of you who have entered into chronic PTSD and who may have also learned to be helpless. While these steps are generally similar to those for acute victims, there are important modifications for victims with longer-standing alterations in more normal biological and psychological functioning, and victims with chronic PTSD will want to pay attention to the differences. The problem of chronicity can be successfully addressed, but it usually entails a longer period of time to master the four basic goals.

The final section on treatment shifts our attention from individual steps to recovery to group methods. Groups, either self-help or professionally-run, are emerging as powerful adjuncts to the individual recovery process. There are groups for combat veterans, persons who have been sexually or physically abused, groups for ACoAs, and so forth. Groups provide victims with different and effective ideas about how to solve problems related to the traumatic event, provide mutual support, and increase self-esteem in the search for a meaningful way to make sense of what has happened.

The steps outlined here for both acute and chronic PTSD are presented in the general sequencing that seems most helpful to the greatest number of victims. There are fairly predictable stages in the recovery process (Herman, 1992), but each recovery program, ultimately, needs to be adapted to particular individual needs. Feel free to modify the sequences as you, your family, your friends, or your counselor may feel is the most helpful way for you, but do complete all of the six steps as each is necessary to help you attain the four basic goals that are central to your full recovery.

The interventions are stated simply, but it will take some time to implement them fully. Take whatever time you need. Have reasonable expectations, do things in small manageable steps, and do not become discouraged. These painful events can be put to rest.

Acute PTSD Episodes

The victim's immediate response to any kind of acute traumatic event is similar to the basic process we spoke of earlier for victims of sexual abuse. The first twenty-four to seventy-two hours requires the victim to respond to the intense disorganization caused by the event. Safety must be assured and medical attention sought, if it is needed. This is followed shortly by attempts to get one's daily life back in some reasonable degree of order. The anger, fear, and sadness that are temporarily put on hold for this readjustment to one's daily routine return in the form of depression and a need to understand what happened. If these issues are left untreated for three months, the victim will move from acute PTSD to chronic PTSD. Here are the steps that can help you address your needs for recovery if you are in the crisis itself or in acute PTSD.

1. Safety First

The first step is to make sure that you are safe. No one can make a successful recovery when he or she continues to be abused. Each episode of violence re-traumatizes you, and further worsens the considerable biological and psychological disruptions that we spoke of at length in the first part of this book.

You must be physically safe first, and physically leaving the unsafe environment needs to be addressed. Should you run for safety? Should you call others for help, such as your family, the neighbors, the police? Each case must be decided individually, but physical safety is a sine qua non. You must be safe.

Seeking medical attention is next. Every victim deserves good care, and you should seek it. Medical attention may include help for any physical injuries sustained; tests for venereal disease, pregnancy, or AIDS in cases of sexual abuse; and possible psychological counseling. Hospitals and emergency rooms can be fairly intimidating places, and even though medical personnel are well-intentioned, some of their routine procedures and detailed history-taking can feel like a second episode of violence. Thus, it is best for you or a loved one to state clearly when you first arrive at the hospital that you are a victim and are feeling totally vulnerable. Family or friends can be designated to give tactful reminders of this to staff in the course of the hospital visit.

Legal assistance may also be sought during these early hours. You may want to involve the police in seeking out the assailant. You and your loved ones may also want to consider the role of legal redress in the short term (such as securing a restraining order in the case of domestic violence), or obtaining counsel and victim assistance services from the court during the course of a later trial.

There are some other practical issues that you need to be mindful of to ensure your safety. Avoid driving or operating heavy machinery, because you may have problems with sudden startle reactions, with concentration, and with intrusive memories, including sudden flashbacks. All of these factors may temporarily impair your coordination. There are other practical concerns as well. If your house was broken into during the crime, does it need to be repaired? Did the street crime involve theft, and has it left you without money to return home? Will the violence draw media attention? If so, how would you cope with this to avoid being re-traumatized when asked to recount the traumatic event in vivid detail? Such issues need to be addressed as part of the overall plan for physical safety, and taking time to do this will greatly decrease your physical distress as well as augment your sense of being in control.

The second component of safety is safety with trusted others. Trust is especially important if you are to discuss the traumatic event. When a victim has decided to disclose what has happened and to share the nature of the violence, he or she should be prepared for the whole range of human reaction from acceptance and support to anger, disbelief, and even victim-blaming. This is especially true in cases of rape, incest, and domestic battering, and this is why it is important for you to select someone whom you can truly trust. Medical personnel, mental health counselors, clergy, police, and lawyers can often be helpful. Other victims certainly understand such things. Family, loved ones, neighbors, and the larger community may be initially shocked that such an event has happened to someone they know, but in time most families become supportive of their loved one. You want to select your trusted person wisely. In the midst of the turmoil, the trusted person will be able to help you find a safe haven where you will no longer be victimized.

But how does any of us know that we can trust someone? What is trust? Is it a feeling or an attitude? Should we blindly trust someone until they disappoint us? Can you trust someone when you first meet them? There are actually useful guidelines to help us in such matters. Trust is not guesswork. The assessment of trust comes from your interaction with others, and is a function of observable behaviors and known values.

Trust is composed of predictable behavior and prosocial values. Predictable behavior means what a person says he or she will do, he or she in fact does most times. If your date says he or she will pick you up Saturday at seven in the evening, he or she should be there on time. If you are asked out again several times, and at seven in the evening each time, you should be able to predict the other person's behavior. What happens if your date comes an hour late one evening? You listen for the person's explanation. Is it reasonable? Is he or she on time the next weekend? Since the world is so complicated and often beyond our control, everyone is given some margin of error. If you are repeatedly unable to exercise some reasonable degree of predicting another person's behavior, then you will usually be unable to fully trust that person.

The second component of trust is similar values. Are you and the other person in common agreement on the basic values in life? Do you both believe people should be treated with respect? Do you both believe that it is good to help others? Are you both motivated by concern for the general welfare of all persons? If you can generally predict the other person's behavior *and* you have fairly similar basic values, then you reasonably assume that you can trust that other person.

Sometimes you can predict a person's behavior, but you do not share similar prosocial values. One might tell you he/she will continue to beat you or take your money or drive to endanger. We can predict the behavior because they will do exactly what they say they will do. However, we cannot trust such a person because his or her values are at such variance with our own.

Victims understandably have great concerns about trusting others. Again, this need not be guesswork. You need to interact with a person or observe them from a distance to see if the behavior is predictable and if the values are similar. Choose someone whom you can trust to use such personal information discreetly, if at all. When you have identified someone you can trust, you can reach out for help in dealing with the episode of violence.

2. No Substance Use or Addictive Behavior

With our new awareness of the repetition compulsion and endorphin opiate-like withdrawal and their links to addictive behaviors, it is clear that all victims need to avoid the misuse and abuse of alcohol, drugs, and prescribed medicines. These addictive behaviors complicate recovery because these substances alter the already fragile brain chemistry present in the aftermath of traumatic events. For example, a victim in crisis who turns to alcohol is at increased risk for impaired reasoning, difficulties with memory, and depression. No victim wants to further compromise his or her capacity to respond effectively to the crisis at hand. Trauma plus substance abuse equals two problems to be solved. Each makes the other worse.

Similarly, the other forms of addictive behavior, such as the sexual addictions, repetitive self-mutilation and repeated sensation-seeking may equally compromise the victim's capacity to cope with violence. While brain chemistry may not be as disrupted as it is with alcohol or drugs, all addictions consume great amounts of the person's time as the addiction is pursued. This is valuable time that prevents the learning of better and more effective ways to cope, and delays the onset of recovery.

Do not use addictive behaviors as a solution to your PTSD distress. Do not let your attention be drawn away from the true and painful matter of the violence. The addiction will only provide you with very short-term relief, and the impact of the traumatic event will still be there. If you already had an addiction or incurred one as a result of your traumatic event, get help in this matter. Consider joining Alcoholics Anonymous, Narcotics Anonymous, Overeaters Anonymous, or any of the other fine self-help programs that are available. Professional counseling for addictive behavior may also be helpful. Plan to use the more adaptive ways to seek relief from the physiology of traumatic abuse that are listed in step 5 below.

3. Restore Reasonable Mastery

We have seen how overwhelming traumatic violence disrupts your sense of personal control, and it is important for you to restore some sense of mastery as quickly as possible. When you are safe, there are immediate opportunities to do this.

You can, and should be encouraged to take as much responsibility as possible for seeing that your own needs are met. Medical care; a safe place to stay; needs for clothing or money; notifying one's employer, family, and friends; debriefing medical and police personnel who may be assisting; all are opportunities for you to reassert control.

Likewise, being actively involved in planning your recovery process is another opportunity for you to establish some sense of mastery. It is helpful in these early post-trauma hours to learn about the symptoms of traumatic stress, the feelings that are commonly found in victims, how intrusions may emerge, and the importance of being with others. Such education about the potential fall-out from traumatic events can provide you with an additional sense of control during the hours after the violent episode.

A third way to restore mastery is for you to exercise control over events that have no immediate relationship to the violence. Balancing the checking account, getting the car repaired, running household errands will all help you to restore a true sense of being in control in the early days after the incident.

A fourth way to regain some sense of mastery is to avoid negative thinking, such as excessive self-blame. We noted earlier how faulty overgeneralizations (Dr. Beck) and false assumptions (Dr. Ellis) can lead to increased distress. Some common overgeneralizations in victims include all-or-nothing thinking ("Because I am a victim, my life is forever ruined," "Since I was attacked once, I can never be safe again"), and a discounting of the positive behavior ("I'm safe, but I should have had more presence of mind.") are not helpful, and you should be careful to avoid them.

Equally unhelpful are the faulty assumptions that may be universally found in victims. Common false statements include: "I am worthless because this has happened," "I am defective, unclean, and not loveable," "It is horrible that life has treated me this way. Why go on?" "I am a bad person," "If people really knew me, they would be repulsed," "I am fat and ugly," (or some other disparaging characteristic), "I am a failure," "I can never do anything right," "I must be going crazy." Statements such as these can destroy your self-esteem. While self-blame may provide some initial sense of control, over time these unchallenged negative statements may result in helplessness or the catastrophic thinking that can lead to possible panic attacks. It is best to avoid them as much as you can.

One helpful way to deal with negative thinking is talk to yourself in positive statements. With your trusted person, write out on index cards five positive statements that you can truly make about yourself (let your trusted person be the judge). When you begin to think negatively, shift this unhelpful mind-set by reading your positive statements. Counselor Rochelle Lerner (1985) has some affirmations for victims, if you want some help getting started.

A final way to restore a sense of mastery is to use the six characteristics of stress-resistant persons that we noted in chapter 2, as you attempt to cope with the trauma. I have developed a program to teach these skills to victims who may not have mastered them as yet. The program is called Project SMART (Stress Management And Relaxation Training) (Flannery, 2003), and the specific steps on how to start your own program for yourself or with your friends may be found in appendix A.

4. Maintaining Caring Attachments

We have seen how links to caring others improve the functioning of our cardiac and immune systems, and how such links result in endorphin enhancement. These health benefits are helpful to victims as is the support, companionship, and problem-solving assistance that such relationships offer. They provide victims with a sense of belonging and improve self-esteem.

It is very important for you to fight the urge to withdraw from others. Learn whom you can trust, using the guidelines we outlined above, and then reach out to those safe persons. Families, friends, counselors, recovered victims will all be able to help you organize your life in the trauma's aftermath. Trusted others can serve as a container for the feelings that you have, and give you advice on what to do.

Victims often do not want to talk about what happened. They may be embarrassed, have feelings of shame, or just believe it will not do any good to talk. If you are one of these quiet folks, you may find the research of psychologist James Pennebaker (1990) of interest. He found that talking was good for your heart and your health. When his victims wrote or talked about what had happened to them, they felt better, thought more clearly, their blood pressure was lowered, and their immune systems functioned more effectively. Talking about your feelings of anger, betrayal, and sadness in the presence of trusted others will hasten your recovery.

5. Tone Down the Emergency Mobilization Systems

It is important to remember how our bodies respond to crisis. The heart works harder, muscles tighten up, and vigilance is heightened as the body and brain become ready to respond.

All victims, including you, need to regularly pursue some stragegy to reduce the physical arousal associated with traumatic events. Vigorous aerobic exercises such as jogging, swimming, or brisk walking may be helpful to you in reducing stress arousal and in enhancing endorphin functioning. In addition, avoid excess use of caffeine and nicotine. These strategies treat both the arousal state and/or the opiate-like withdrawal issue and are more effective coping strategies for you than the various addictive behaviors that we have mentioned.

Relaxation exercises may be equally helpful. They can be implemented fairly quickly and can be done unobtrusively. There are many varieties to choose from. Deep breathing or meditation, self-hypnosis, prayer, systematic muscle relaxation, viewing art, listening to soft music, and certain hobbies like photography or knitting can all be helpful ways to reduce the emergency arousal systems once the crisis has passed.

My colleagues and I in the Victims of Violence Program at the Cambridge (Massachusetts) Hospital have taught many victims to relax. Sometimes when victims first try to relax, they feel out of control and frightened. We have modified such relaxation approaches to meet these differing needs of victims so that they can relax but still retain a sense of control. These modifications can be found in appendix B, and they may be helpful to you.

Choose whichever system feels most comfortable to you, but do include some approach to reduce arousal right from the start.

6. Making Meaning of the Violence

Making some meaningful sense of traumatic events requires that you grieve the loss, have some understanding of the origins of violence, and develop a renewed sense of purpose in life. We begin with grieving.

To grieve properly, you need to acknowledge and accept the reality of what has happened. Denying or rationalizing away the event will only postpone the grief work that needs to be done. It will not go away until it is directly dealt with.

You will want to allow yourself to experience the sadness associated with the grief as you go through the stages of denial, anger, bartering, despair, and acceptance that we outlined in chapter 2. It is best if you share the grief with another who should listen quietly, avoid blame-making, and help to foster your expressions of anger and sadness (Kushner, 1981). Counselors may be of special help with this part. Many victims find it helpful to begin with a detailed description of the thoughts, feelings, and specific facts as best they can recall them. As you review what has happened, you may experience feelings of sadness, betrayal, depression, and the range of emotions that humans suffer in the face of loss. Reviewing the experience in detail repeatedly may be helpful for you because with each recounting your thoughts and images are more clearly pieced together, and previously dissociated thoughts and feelings may be recalled. All of this is best done in small gradual steps over time.

Mourning allows you to regain some control over the feelings of fear, anger, guilt, shame, and depression that we spoke of in chapter 2. It allows you also to explore the victim themes noted by Horowitz (1986). Themes of fear of repetition of the event, fear of one's own aggression, and the like are common. You may want to review Horowitz's list (1986) in chapter 2 to see which may be applicable in your own case. Issues of feeling subjugated, unclean, defiled, damaged, and guilty are also common, and you need to share them with your caring person so that he or she can help you correct your distorted perception of what has happened. In reexperiencing the traumatic event in such a grieving process, you relate the event to the self now, and not to the self that was overwhelmed and out of control. It will help you to accept any permanent changes and/or losses in your life.

Some victims of violence have trouble dealing with their anger. Some have been directly threatened with death if they made a sound, others have paired anger with violent death from their own personal experience of the trauma. In many cases, victims have confused normal assertiveness with violent rage. Normal assertiveness is speaking up for your own needs and wishes in a direct, but tactful way that invites the cooperation of the other party. Rageful anger seeks its own way regardless of the rights of others. Victims who are afraid of their anger are not usually rageful people. They are afraid, however, that they may be, and that their anger may actually destroy others. If you feel this way about yourself, it is best to discuss your apprehensions with your trusted person. Anger can also be addressed in small and gradual steps. For those victims with long-standing issues of nonassertiveness that pre-date the traumatic occurrence, self-help groups or short-term professional counseling around this specific issue may prove helpful to you. Anger is often one of the most difficult issues for victims to master, and it may take you some time to feel comfortable with it. As cited earlier, Carol Tavris (1982) has written a thoughtful book about anger, if you want to read further.

As your mourning concludes and your traumatic event becomes reintegrated into your present sense of self, you will want to begin to explore the impact of this event on your basic values. What have you learned about violence from your encounter? What have you learned about yourself? What strengths and weaknesses did you observe in yourself? How would you cope more adaptively, if it should ever happen again? The answers to these questions will help you understand which of your values are most important to you.

With your grieving completed, you are now ready to explore the origins of traumatic violence and to find a renewed sense of purpose and meaning of life. These are complicated tasks and the subject matter of chapter 10.

*   *   *

The completion of these six steps will enable you to resolve your acute traumatic crisis, and to return to more normal living. You will hopefully be wiser for your experience, but no less impaired in your capacity to enjoy life.

Chronic PTSD States

Chronic PTSD states are episodes of abuse that were not treated initially. Victims in the chronic phase need to go through the same six steps as victims of PTSD trauma, but with the specific modifications noted below.

1. Safety First

Victims of chronic PTSD are no less in need of safety than are victims in acute situations. If you are such a victim, you should observe all of the steps noted previously in the safety step for acute victims. This includes the guidelines for trusting others and for disclosure of what has happened to you.

As a victim of chronic PTSD, you may have an important, additional problem. Some of you may be in situations where you continue to remain unsafe. Incest and physical battering are possible examples of this dilemma. If you remain in situations where such repeated violence continues to re-victimize you, full recovery is highly unlikely to occur. In such situations, you should review the guidelines for trusting others. When you are assured that you can trust someone with the secret, tell them what is happening and ask them to think through with you the best and safest means to seek a safe place. They can contact others anonymously (e.g., a lawyer, a safe family member, children's services, a women's resource center, etc.), and get further advice on what to do. Plan your route to safety carefully and follow it. To be sure this is easier said than done, but others have done it before you. Utilize small manageable steps, take your time thinking these matters through, and exercise great care when you implement your plan.

When you are safe, you should be sure to have a physical exam. Neurological and endocrinological exams should also be obtained, if your physician so advises. These exams will be helpful in determining any possible longstanding effects of the abuse. For example, if you were physically abused earlier in life, you may have sustained a head injury that has gone undetected. Good medical exams can hasten recovery.

Victims of chronic PTSD usually have long-standing issues with trusting others, and you should feel free to spend extra time studying the guidelines for trusting others, and observing the behavior of others so that the process of trusting can be gradually restored. Trusting others may be a continuing issue for you to work on.

2. No Substance Use or Addictive Behavior

Substance use and addictive behaviors complicate recovery in the ways we have noted earlier. Since many chronically abused victims have often started to self-medicate with some form of addictive behavior, sobriety or being straight is a must. If you have been addicted for a long time, it can be several months (in the case of alcohol, twelve to eighteen months) before your brain chemistry fully returns to normal. Other forms of addictive behavior besides drugs and alcohol must also be avoided. Concurrent treatment for trauma can begin when your addiction is controlled and your mind is clear. A recovery process is temporarily blocked by an addiction, so it is important to keep trying to master the addictive process. Any Alcoholics Anonymous meeting can show you that it can be done.

3. Restore Reasonable Mastery

As a victim of chronic PTSD you are subject to the same issues of loss of control as are acute victims. You too need to engage in the various steps to restore mastery that we have already noted: take control of some aspect of your life, plan your recovery, avoid the various types of negative thinking, and develop the skills of stress-resistant persons.

Because your trauma has remained untreated for a long period of time, you may have additional issues with mastery and helplessness.

The first problem is that of learned helplessness. As we have noted earlier, this stance of helplessness by its very nature may be keeping you from solving your more fundamental trauma issues. The helplessness must be treated first.

The Project SMART program has proven to be useful in this regard. When the criteria for safety and elimination of addictive behavior have been met, some helpless victims recover much more quickly if they join a Project SMART program first. As noted, the steps of the program have been written out in appendix A. If you have developed helplessness, and participate in a Project SMART group, you will learn the characteristics of stress-resistant persons even as you begin to overcome the helplessness. You will then be able to correct what negative thinking you may have learned over the years, and then take self-initiated action in restoring your life to normal.

The second possible mastery problem for those who are victims of chronic PTSD is the development of panic attacks. We have seen in chapter 2 how such attacks in victims may result from faulty labeling of body events as catastrophic (Hawton et al., 1989).

The best way to reduce panic attacks from trauma is to challenge your automatic faulty thinking. What is the real evidence that you are going crazy? Or are having a fatal heart attack? and so forth. Are you thinking in an all-or-nothing frame of mind? Are you really having an anxiety attack or have you mislabeled normal body processes as catastrophic events? How would someone else view what is happening to you? And what would actually happen if you did have the panic attack? You would be uncomfortable, but you would not go crazy, nor be totally out of control.

When you fear a panic attack, use a form of relaxation, try not to be overcontrolling of the event, challenge your catastrophic thinking, and face the situation with your new perspective. In time, when you feel better, make yourself deal with issues or places that you have avoided for fear of panic attacks. Take a "safe" person with you the first few times, until you learn that nothing catastrophic is going to happen.

A long-standing history of panic attacks can be a nuisance that is hard to master on one's own, because often most every place comes to signify a possible panic attack. If you would find a counselor to be of help, there are very effective professional treatments for panic attacks. Your family physician or local health care agency should be able to provide you with a list of qualified people who can be of assistance.

4. Maintaining Caring Attachments

Victims of chronic PTSD often withdraw from the network of caring attachments that was theirs. The need for caring links to others is no less true for chronic abuse victims than for victims in acute PTSD.

Review the steps for trusting others, and some of the suggestions that were offered earlier in step 4 for victims in the acute PTSD phase. Start in small ways. You need only to build or rebuild one attachment at a time. Rely on others in little ways at first. Gradually you will feel more safe, more relaxed, and more supported by others. Building links to others often takes longer for victims of chronic abuse, so be patient with yourself in the process.

Some victims of sexual abuse may additionally experience issues in restoring physical and psychological sexual intimacy. Counselors can be helpful with these matters, and family therapist Dr. Gina Ogden (1990) has written a useful book for victims on this special topic.

5. Tone Down the Emergency Mobilization System

As a victim of longstanding abuse, you are more likely to have problems with kindling and its opiate-like withdrawal, and you should include a method of relaxation in your recovery program from the list of methods that was presented earlier. Aerobic exercises (except for victims with panic attacks) and relaxation exercises modified for victims (see appendix B) may be particularly helpful for chronic arousal states. Avoid excess caffeine or nicotine. As a victim with long-standing untreated states of chronic arousal, you will need to practice consistently to maintain low states of arousal.

Medicine is sometimes helpful in reducing the long-standing symptoms associated with chronic PTSD. Antianxiety medicines can be helpful for hypervigilance, sleep disturbance, and some of the other symptoms associated with physiological arousal and avoidance. Antidepressant medications can be helpful with the intrusive memories, including flashbacks, as well as panic attacks. Seek the advice of your physician in these matters if you feel medicine may be helpful. Since addictive behavior is common in chronic PTSD victims, you and your physician will want to monitor the prescribed medicines to avoid any excessive use of what your doctor has ordered. If medicines are used correctly, you need not unnecessarily fear becoming addicted again. This is a personal issue, however, and each victim must decide this matter for him- or herself.

6. Making Meaning of the Violence

To make meaningful sense of traumatic events, victims of chronic

PTSD like victims of acute PTSD need to grieve the losses in their lives first. Exploring the roots of violence and searching for a new purposeful meaning in life come later.

To grieve properly, you need to go through the same steps as we noted earlier: accept the reality of the event, experience the grief, mourn the loss, accept the new limitations, and review basic values. (See step 6 in acute situations for greater detail.)

There is one important difference, however, in cases of long-standing abuse. *The traumatic event must be grieved in small manageable steps.* This seems to fly in the face of common sense. Most victims, when they finally decide to face the event, want to do it all at once, and be done with it. *Recalling the traumatic event in vivid detail with all the painful feelings all at once is not helpful. You may be biologically re-traumatized and once again feel psychologically overwhelmed. Everything may be made worse and your recovery may be delayed. The basic strategy for grieving abuse of a chronic nature is to recall the traumatic event in the grieving process with your trusted other or counselor in small manageable steps so that each aspect of the event can be mourned, and so that its painful feelings and memories are gradually assimilated into your present self.*

For example, weekly steps might start with a discussion of where the event took place. A later discussion could include when the assailant was first noticed. This could be followed in the third week by when the victim knew he or she would be victimized. A discussion of a part of the actual trauma would follow next, and so forth, until the whole episode is recounted. This is a painful process, but it allows you to assimilate the event and any dissociated aspect of the event, fully. It allows you to experience the fear, anger, shame, guilt, and depression common to the event, and in time, to gain mastery over the event.

Since the recall of past traumatic events can be overwhelming for some victims, a brief hospitalization may prove helpful in cases where the victim feels acutely suicidal, becomes severely depressed, or needs help in containing his or her feelings. Trauma is a medical condition, and no victim should be ashamed of seeking any necessary medical care.

The medical research is unclear as to whether flashbacks and intrusive memories can ever be fully eliminated, but these six steps can markedly reduce their frequency and intensity. The re-search is also unclear whether every victim can recall past trau-matic events in a way that is helpful to him or her. Some victims appear so overwhelmed by the recall of such episodes that in some cases it is better to leave the event sealed in memory and to work on the other steps toward a more normal life. This would leave the victim subject to possible intrusive memories in the face of threat or loss, but given our current understanding of treating these events, this may be a better solution in some few victims. Each case needs to be decided on its own merits. Since victims often want to avoid dealing with such events and memories, if you are wondering what is best for you, ask a counselor, another victim, your family, or your physician for advice.

As with victims of acute PTSD events, when you have com-pleted your grieving process, you are now ready to explore the origins of violence and to find a new meaningful purpose in life, tasks that are addressed in chapter 10.

*   *   *

The completion of these six steps will enable you to attain a more normal daily routine, and to be freed of the burden of untreated trauma that so hardens the heart. Some of you may continue to recall the event, but you will now have better control of such intrusive memories and the painful feelings that they evoke. In following these steps you will enhance your stress-resistance and attain the treatment goals of reasonable mastery, caring attach-ments, and begin the process of finding a meaningful purpose in life in your newly relaxed state.

Group Treatment

Groups can be very helpful additional tools for recovery for victims, in addition to victims' individual efforts to resolve such events.

Groups provide any of us with forums for acceptance. This is especially true for PTSD if the group is composed of other vic-tims. As we noted earlier, other victims can facilitate the new member's disclosure of the violence, and can provide new ideas about how to cope with the aftermath. Such help generally precludes the development of helplessness, and restores the victim's sense of mastery.

Groups also provide a context for the development of caring attachments. Such attachments help victims to avoid shame, guilt, self-blame, and feelings of vulnerability. Groups generally focus on trust, and helping one another, and concern for the other group members provides an early beginning for a renewed sense of purpose in life.

Groups facilitate recovery and are found in two general formats: self-help groups, and professionally-led counseling groups.

In 1987, over six million Americans seventeen years of age or older belonged to some type of self-help group. There are at least two hundred different types of such groups for medical problems, the various addictions, and the range of differing traumatic events.

Such groups are very democratic. They are member-run and governed. Every member has equal rights and each member has personal experience with the main purpose for the group's existence. Ask around about such groups, read the calendar of events in your local newspaper for a listing, or check the white pages. The fee for membership is nominal, if there is any at all.

Professionally led groups can be equally helpful. In the Victims of Violence Program at the Cambridge (Massachusetts) Hospital, my colleagues have developed a three-tiered approach for the needs of sexually and physically abused victims (Koss and Harvey, 1991).

One group provides victims with support and education about what trauma is and what its untreated consequences may be. The group focuses on trust, control, support, and initial feelings of loss, anger, and guilt. A second group format consists of Project SMART groups (see appendix B) to help victims regain reasonable mastery, develop caring attachments, and avoid helplessness. A third group level consists of concentrated work on the traumatic event itself. Traumatic events are shared, memories are recovered, and some direct control of the traumatic experience is encouraged. Goals here might include sharing the secret, improving relationships, thinking better of one's self and so forth.

Each of these three group formats are short-term interventions lasting no more than ten weeks each. Victims begin in the group most suitable for their initial needs. These groups are proving to be helpful adjuncts for individual recovery (Koss & Harvey, 1991). For victims with continuing interests in and needs for longer-term groups, most mental health agencies provide continuing survivor support groups.

Some victims are initially frightened and appalled by the thoughts of joining a group. This is understandable, as victims want to withdraw and do not want to be mistreated by others again but many victims are greatly relieved and pleasantly surprised when they see how helpful such groups can be. Groups are not for everyone, but I would again encourage you to see if a group would be helpful in your individual case.

*   *   *

This chapter has focused on the six steps that victims may choose to utilize to resolve the aftermath of psychological trauma: (1) safety first, (2) no substance use or addictive behavior, (3) restoring reasonable mastery, (4) maintaining caring attachments, (5) toning down the emergency mobilization systems, and (6) beginning to make meaning of the violence to lead to the four goals for recovery. With time and patience you will attain the sense of inner peace that is rightfully yours.

You need not do this alone. You can do what stress-resistant people do: reach out to others.

As we have noted repeatedly, caring attachments can be powerful resources for helping victims to recover. Not only are such attachments found in groups with other victims, but they are all around us if we care to utilize them. We shall turn now to specific ways in which your family and friends, counselors, and the legal system can be helpful components in your recovery program.

9

Recovery:
Help from Family and Friends

Human beings rust from within
from the tears never shed.
— Anonymous

Stay with us for it is getting toward evening.
— St. Luke

"When I tell you to pick up your toys, pick up your toys."
Little Cheryl started crying. She was only four.

Russell Irving was late for work. He'd hug his wife and daughter when he got home. The morning had brought a heavy snowfall to New York City. Fire fighting was never easy, and certainly not in snow. He could predict that response time to calls would be longer, that ladders covered with ice would make everything more dangerous. What he could not predict was that his life would never be the same after today.

"Flight 741 cleared for final approach. Descend to 5200."
"Roger, Tower. Flight 741 to 5200." "Flight 741 descend to 5000."
.... "Flight 741, do you read me?" .... "Flight 741, this is the tower, do you read me?" Flight 741 had fallen from the sky, and had crashed into the earth in one hundred thousand shattered pieces.

NYFD Engine Six raced in the snow through the Borough of Queens. Russell sat in the cab and picked absentmindedly at his nails as he mentally reviewed the procedures for disasters.

The scene was surreal. A huge ball of fire in the midst of blinding snow in one of the largest cities on earth. The stench of jet fuel. The cries of survivors. Thick acrid smoke. Aircraft parts. Human body parts. Fire everywhere.

The men of Engine Six searched for survivors first. Russell found a couple holding hands and still strapped in their seats. Both were dead. He saw a woman's arm on the ground before him, and the dormant body of a clergyman. He was nauseated, but he steeled himself to continue the search. He saw a little girl in the bushes near his feet. Russ started CPR. As she responded, he picked up the child and raced her back to the EMTs near his truck.

As the EMTs fought vainly to save her, the local television news crew swarmed around him for an interview. He turned away in disbelief, and resumed his search for the walking wounded in this technological holocaust that engulfed them all.

Thirty-six hours later, Russell Irving left the station, and drove home in the darkness on the now cleanly plowed streets. He parked the car in the driveway. Walking on tiptoe, still in his heavy firefighter boots, he went to his daughter's bedroom, enfolded his little Cheryl in his arms, and fell to his knees in tears.

*   *   *

While the last chapter focused on individual strategies that you as a victim could implement on your own, this chapter focuses on resources that are available through the efforts of others. There are three major resource networks that you may want to consider. These include family and friends, health care providers and emergency services personnel such as the New York Fire Department in our chapter example, and the judicial system. Each of these groupings of individuals has special skills that can hasten full recovery. You may wish to carefully consider implementing them fully in your own treatment plan for recovery.

Families that are good problem solvers and are cohesive can be very helpful resources to you as a victim. Both by direct help and by enlisting the special support of others such as a family doctor, families can do much to help you disclose the traumatic event, and they provide you with assistance in resolving the problem. If the assailant is a family member, then your family will need outside assistance, and we shall consider this special case later on. Otherwise, your family is often a first potential resource network that you will want to consider.

Another helpful network may be found in the assistance of health care workers and emergency services personnel. These support networks are automatically activated when a victim comes to an emergency room, when natural or man-made disasters strike, or whan a victim seeks professional counseling. These expert trauma specialists have had long hours of training to be there when you need them, and, as a victim, you should avail yourself of their skills.

Finally, the remaining support network available to you is the legal system. The police, courts, and victims assistance programs exist to help provide redress to victims. It is helpful to know of these services and to be able to use them if you wish.

We live in a society that is increasingly violent. Untreated victims and those in the recovery process remain at increased risk for revictimization in a violent society. For these reasons, I have included at the end of the chapter some basic steps for prevention during the recovery process and for general safety in today's age.

As with the strategies for victims in the last chapter, you, your family, and health care providers will need to adapt the general suggestions presented here to the particulars of any individual situation. With this in mind, let us review the basic resources in our society that are generally available to all victims. We begin with families.

Help From Family

The first potential major resource for any victim is the victim's own family. Healthy families are cohesive and good at problem-solving, when things go wrong. Psychological trauma is one of those times. Your family can aid in helping you reestablish some sense of control, can help combat your painful feelings, and can help you figure out the best path to recovery. They will rally to your support in search of constructive solutions.

But they cannot help you if they do not know what has happened to you. Not disclosing the episode of violence will affect every family member. If the crisis in your life is a recent event, your family will know that something is wrong. They will sense that you are troubled or frightened, and everyone will remain ill-at-ease to some degree, until you have shared your secret and received some treatment.

Medical research has also demonstrated long-term effects on family life when the traumatic event has not been shared and treatment sought (Figley, 1989; Krugman, 1987). Families with untreated victims appear to have more than average hostility and arguing. The trauma victim, in part because of the biology of chronic arousal, and in part because of continuing anger at being defiled may engage in repeated acts of verbal and physical abuse to other family members. Perhaps the victim feels that the family will be safe, tolerant, and understanding; perhaps the rage is just too hard to contain. Whatever the reasons, families with trauma victims may be more prone to anger and even to violence, and family members may become understandably frightened, angry, and avoidant of the family member who is the victim.

A second characteristic that these families may develop over time is very poor skills at intimacy. I am referring here not just to sexuality, but to the whole range of interpersonal skills that results in the caring attachments noted in chapter 2. Victims avoid emotional topics, are very reticent to share much about how they feel, and essentially withdraw from the family emotionally. Over time, such poor communication leads to lack of trust, insecurity, and continued feelings of unhappiness even in the best of families.

A third possible consequence may be a theme of general unhappiness and dissatisfaction with life. The normal process of caring attachments that provide companionship and support does not appear to flourish in families with untreated victims. Victims are hard to cheer up, and are prone to repeated recurrences of crises. Such problems as substance abuse, sexual dissatisfaction, and faulty child-rearing result in a sustained state of discontent for all members of the family.

A fourth possible outcome in families where victims remain untreated for a long period of time is the development of co-dependency in other family members. As we learned in chapter 7, codependency is an unhealthy dependence on one person, to the point where the codependent sacrifices his or her needs and sense of self for the needs and wishes of the other. Such codependency can occur in homes with untreated victims. This is especially true if the untreated victim is engaged in addictive behavior. The codependent may feel sorry for the victim and may end up facilitating the addiction. Codependents do this by keeping the addiction a secret, looking the other way, providing money for the addictive behavior, or helping to keep the victim out of jail.

Family as Victim.   There are some cases where your family may not be the beneficial source of help that you would like. Such cases arise when the entire family is victimized at the same time by the same event.

Natural and man-made disasters are some examples of how this might happen. Fire, flood, hurricanes, nuclear accidents, and the like dislodge every family member, and even whole communities. In these cases, families themselves need to rely on others for support and recovery.

The homicide of a family member is another form of traumatic event that affects an entire family all at once. Negligent motor vehicle homicide, murder of political prisoners, violent death in the aftermath of crime may leave even the most stress-resistant family immobilized.

The circumstances of these crimes can further compound the family's pain. Witnessing the event, knowing the assailant, hearing of the death through the media rather than being personally notified may intensify the shock. Violent deaths accompanied by sexual assault or dismemberment of the deceased's body increase the probability that family members will feel even more vulnerable and helpless.

In these cases the family should understand itself to be a true victim. Family members should reach out for help immediately to family and friends, clergy, counselors, crisis intervention teams, or other social service agencies. Even the most adaptive families need help at times like this.

What the Family Can Do to Help.   When your family learns about the violence that has befallen you, you should assume that they may have all of the normal human feelings that are stirred in any of us when we are confronted by such destructive acts. They will be distressed for you, and, to a lesser extent, for themselves. They may have feelings of outrage, disgust, and grief at what has happened to you, their loved one. Moreover, they may also experience fear and depression for themselves, since violence teaches us how fragile life is.

These family reactions may further scare an already frightened victim. It is helpful for you, the victim, to remember that such feelings are normal in caring families, and that they are not directed at you personally, but at the maliciousness of the assailant. These feelings by your family may actually better help you to understand your own experience of the event. In any case, these strong feelings will diminish, but the energy that has been released will mobilize everyone to find a helpful solution for your needs.

Listed below are some specific ways in which your family can be helpful to you in the midst of the turmoil. If you are the kind of person who finds it difficult to ask directly for help, you may want to leave this book in a prominent place so that one of your family members can read it, and hopefully understand your request for help. Here is what families can do:

1. Family members may be particularly skilled at helping you disclose the pain that has occurred. They will know that something is not right, and with supportive caring they may make it easier for you to share what has happened. A colleague of mine at Harvard Medical School, psychiatrist Leston Havens (1986), has developed an interesting approach that he calls performatives. Performatives are statements that perform an action simply by being spoken (e.g., "I christen this ship 'The Mayflower'"). When some aspect of a victim's sense of self is under heavy attack from self-blame or social censure, performatives may prove helpful in supporting that part of the self. Empathic performatives such as "I hope someone you know and trust will put his or her arm around you" or "You bear your pain with great dignity" may lead victims to seek comfort in the first case, and to admire themselves in the second one. Performatives are complex and subtle, and you may want to read further (Havens, 1986). Such statements by parents, however, may be of great help to you in eventually disclosing your episode of violence.

When you have been able to disclose the event, your family can help in other basic ways. They can make sure that everyone is now safe, they can encourage you to express more fully your thoughts and feelings about what has happened to you, they can comfort and support you, and they can see to it that you receive any medical care that may be necessary. They can also help you avoid withdrawing from the family, keep you from assuming that you are helpless, and help you avoid addictive behaviors as a method of self-medication.

2. Wise families will help you to remain as calm as possible in the midst of the crisis by encouraging you to go for walks, engage in aerobic exercises, and/or use relaxation techniques to dampen down the physiology of arousal. Many families find it helpful to do these exercises with the victims as it helps everyone remain calm, and also provides companionship for the victim.

3. Next, your family can help you restore some sense of mastery and control. They can involve you in your own recovery immediately, and you and they can use the information at hand to develop a recovery plan. This plan may include counseling, legal help, and so forth. Who will go to the doctor with you? Who will call the family lawyer? Who will let your boss know that you will be out for a few days? These are the types of questions that families can be helpful with. In addition, your family can be very helpful to you if they exercise some reasonable control over any aggressive outbursts you may have. There is a helpful balance between supporting for your recovery and setting limits on your behavior that is potentially harmful to you and to other family members. Wise families monitor both.

4. You will need caring attachments during this difficult period, and your family members are the resident experts in knowing what your needs may be. Families can provide the companionship, emotional support, and instrumental help that we spoke of in chapter 2. You need to hear that you are safe, that the act of violence was not your fault, that you have not been somehow permanently defiled, and that you remain loveable. Your family can be very helpful to you in repeatedly reminding you of these things.

5. Finally, your family can initiate the process of helping you to find some renewed purpose in life. They can help you reflect on what has happened, why you responded as you did, how such

events can be put behind you, and what could be done differently the next time, should such violence ever happen again (Figley, 1989).

As your family helps you through your crisis in the manner outlined above, they will also be increasing your capacity to cope effectively, because these suggestions, like the individual steps in the last chapter, all result in the adaptive problem-solving skills of stress-resistant persons.

Here is an illustrative example of how this adaptive process might work in a caring family:

Mrs. Harrison had recently noted money missing from her wallet. After discreetly watching her purse for a few days, she realized her eight-year-old son John was taking the money. She feared he was using the money to buy drugs or alcohol. In a supportive but firm way, she asked John why he was taking the money. John at first became angry and denied it, but he then quickly broke down into tears.

He related with sheer terror how an older man in their town house complex was insisting each week that John pay five dollars for the "privilege" of performing oral sex on this man and two of his friends or face being assaulted by them.

Helen Harrison comforted her son and called her husband Carl at work. Both parents reviewed the facts with John. Carl went to the police to seek assistance, and Helen spoke to the school principal about obtaining some counseling for John.

Here is an example of a healthy family using reasonable mastery and caring attachments to begin to resolve a major crisis for one of its members, as it seeks the necessary help to resolve the violence and to restore a meaningful purpose to life.

What to Do When a Family Member Is the Assailant

Sadly, as we have seen, in some families the assailant is a member of the family itself. Betrayal by those who should love us makes traumatic violence even more painful, and, by definition, dysfunctional families of this sort are not the cohesive families that are good at being helpful to one another. Incest, domestic violence, and child abuse are common examples of family breakdown, and they have a crushing impact on everyone.

In such homes, some members of the family are literally unsafe. The parents, the older siblings, the spouse, someone has become the enemy. Those whose duty it is to protect and care have themselves lost control. Victims in such families live in constant fear of the violence being repeated, and feel equally betrayed and angered by adult family members who have permitted such things to continue.

Normal daily life is disrupted as the victims' energies are focused on protecting themselves. Tasks at work, in school, or at home, such as rearing children are impaired. The social development of such victims may be equally arrested as the focus for survival consumes much time, energy, and thought. Other family members may fear for their own safety, and often a codependent member is enabling the family chaos to continue.

Codependency or not, when the assailant is a family member, normal family rules are absent. We have seen the legacy of such behavior on families in chapters 5, 6, and 7. As we noted then, some families become involved only in each other. Life so revolves around other family members that there is no contact with others, no chance for feedback and learning to occur. Other families become disengaged from one another. The only way to cope and to survive is to withdraw physically and emotionally, so that each member has his or her own space. Still other families reflect true chaos. Behavior is unpredictable, erratic feelings punctuate life, and planned orderly thought is rare. In each of these three situations, roles and responsibilities are blurred as each family uses denial and avoidance to coerce silence and maintain the secret of intrafamily violence.

When the assailant is a family member, the family as a resource for the victim is much more problematic. Feelings of mistrust, betrayal, and rage bring any semblance of normal family support to a full halt. If you are in such a painful situation, you and your family will most likely not be able to resolve this on your own, and you and/or your family should seek outside assistance.

Counselors can be helpful at such times. The counselor is there to help reduce the family's sense of being out of control. Safety is the issue that will be addressed first to ensure that each and every family member is safe. The counselor might ask the offending family member to temporarily leave the home, restraining orders might be obtained, care and protection orders for the children could be secured. Each of these possible steps by the counselor is completed as a caring act for all of the family members, and such acts are the first step in restoring family normalcy.

Most counselors of families focus next on establishing the roles and responsibilities of each family member. The adults will be expected to provide care and protection, and will be educated in how to accomplish these goals. For example, we have seen, in father-daughter incest cases, how the mothers in these homes are passively uninvolved with their own needs and the needs of their children. The family's counselor would work with the mother to help her meet her own personal needs and to become more assertive with her children.

The third focus of the family's counselor is to discuss slowly the traumatic incident(s). This happens only after normal family functioning has been restored, and the family members feel less crisis-ridden. The counselor will review the painful episode, help the victim begin to reintegrate what has happened, and plan with the family a longer-term full recovery process for the victim and the family so that such events will not recur in the future. As time passes, the family will feel more in control and have a restored sense of hope as more normal living begins to emerge. At this point, the role of the counselor diminishes.

Here is an example of a family situation in which the assailant is a specific family member:

Grace was very concerned about her niece Rose. Rose's grades had slipped badly in the past year, and this was the third time she had run away from home. Grace knew that the early teen years were difficult for any youngster, but something seemed wrong, radically wrong with Rose.

When the police brought Rose home for the third time, Grace was startled to hear from Rose that her stepfather had been sexually abusing her for seven months, and that Rose's mother knew about it, but had done nothing.

Grace did not know what to do either, but a call to her clergyman got her the number of a local rape crisis intervention center. Grace and Rose went together that day for help.

Rose was allowed to live temporarily with her Aunt Grace. A restraining order was obtained to keep Rose's stepfather away from her, and both her parents agreed to counseling as an alternative to any further legal proceedings. The parental counseling proceeded in a way similar to what we have just outlined above.

Rose was fortunate in that she had an aunt outside the family whom she trusted and could turn to for help. Not all victims have this opportunity. However, even in cases where there is no family member or neighbor to help, victims still have resources they can trust. Victims can and should utilize the clergy, the local mental health center or other social service agency, shelters, services for runaway teens, and always the emergency room of a local general hospital.

The Intergenerational Transfer of Violence

A question commonly asked by victims of violence is this: Will I too become violent toward others? Will I hurt my family members or others that I come upon? This is an important question for many victims, and you will feel at ease when you learn of the most recent findings.

The intergenerational transfer of violence theory in cases of child abuse states that men and women who were abused as children are considered highly likely to be abusive themselves as adults toward their own children, spouses, or elderly parents. This phenomenon is also known as the "violence begets violence" theory.

The theory of this intergenerational transfer of violence appears logical. People model and imitate what they have learned. The grandparents battered their children, who battered the grandchildren, and so forth. Each generation learns to be violent in adult life by participating in family violence as children, and there are many studies that report a link between childhood abuse and neglect and adult domestic violence.

So are we to assume that the intergenerational theory of violence is correct? Is it widespread? How frightened should you be of repeating a history of violence?

In a recent and excellent review, psychologist Carol Widom (1989) has been most helpful in answering some of these questions. Dr. Widom reviewed the published data about the long-term effects of child abuse and neglect on subsequent violent behavior. Here is some of what she found.

Most children who were abused were not abused by a parent with a past abuse history. In about one-third of the cases the abusing parents had themselves been abused. While this is not acceptable behavior, it also reveals that being abused as a child does not mean that you will necessarily be abusive as an adult.

Dr. Widom also found that about twenty percent of abused and neglected children become delinquent, that there was no clear evidence that being abused as a child led to other forms of violence as adults, but that there was an increased likelihood of such abused children being self-destructive as adults (e.g., self-mutilation and suicide attempts).

Her research review also pointed out that adults with a past history of childhood abuse responded in other ways that were not repetitive of violence toward others. Depression, withdrawal, and various forms of self-punishment were common themes in some.

What health care providers are learning is that there is no simple relationship between childhood abuse and adult violence. Abuse may lead to intergenerational violence, but in many cases it does not. Past and current life stress affects the decision to turn to violence at a particular point in time, and the absence of reasonable mastery, caring attachments, and a meaningful purpose in life may increase the likelihood of violence. The victim cannot erase his or her past history of abuse, but he or she can control life stress, and can develop or augment these three basic skills related to good physical and mental health. In sum, victims need not live in undue fear of harming others violently. To use violence is a choice that is not predetermined, and one that need not be selected as a solution to life's problems.

Help from Counselors

Individual Counseling

Some of you reading this book may wish to reach out for individual counseling, the second of the three major resources for victims. Nonetheless, you may be understandably unsure of when to do this, what to expect, and how you would go about selecting the right counselor for yourself.

A good general rule of thumb is that you should seek extra help whenever you feel you have the need for it. Perhaps you have come to realize that you had issues in coping with general life stress before your episode of violence, and need some help in sorting out the two. Some of you may feel the need of additional support in the recovery process, but feel shy about asking for help. Others may want specific advice, for example, on how to disclose the violent event to a particular person. Some may want to see if medicines can be helpful. Others may seek specific religious counseling to help with the problem of human destructiveness. There are as many reasons for individual counseling as there are individual victims seeking help.

Each counselor's personal style and formal training will be different, and there are many different and effective counseling approaches. You can expect to be listened to thoughtfully, to be supported in your efforts to overcome the impact of violence, and to be helped to plan strategies for your own recovery. It is the counselor's goal to help you to return to full independent living, and in addition to the individual counseling, the counselor can put you in touch with professionally-led group treatments and other community resources such as child welfare services, women's resource centers, support groups, various national "800" hotlines, and the like.

Choosing a counselor need not feel overwhelming. Ask other persons who have sought counseling whom they would recommend, or ask your family physician. When you first meet with the counselor, ask him or her about his or her professional training, and be sure that the counselor has some experience counseling trauma victims. Nonmedical counselors need to have links to the medical community, and you should ask about such matters.

You may want to visit two or three different counselors until you find one with whom you feel most comfortable. Remember that the counseling will raise painful issues for you, and you must remind yourself not to drop out of treatment when the going gets tough.

Any form of intrusive abuse on the part of the counselor such as sexual advances, sharing drugs or alcohol with you, or seeing you outside of the counseling relationship is unethical and wrong. This does not happen often, but if it should happen to you, stop immediately and find another counselor.

Community Counseling

Many of us do not realize that there is a whole network of emergency services personnel who are available to help us when

whole communities are victims of violence through natural disasters like volcanoes, hurricanes, and earthquakes, or by means
of man-made disasters such as the collapse of buildings, rail and
air disasters, or toxic chemical spills. Such personnel include fire,
police, emergency medical technicians, Red Cross volunteers,
emergency room staff, mental health personnel, and civil defense
agencies.

Emergency services personnel are highly qualified men and
women who have spent many long hours in training to respond
exactly to the disaster you and your neighbors may find your-
selves involved with. Within twelve hours after disaster strikes,
all normal semblance of community life is disrupted. Emergency
services personnel know this, and step in quickly to begin to help
put the pieces back together.

When some degree of community order has been restored,
mental health personnel work with the victims to treat the psychological aspects associated with such trauma. This is most often done through a process known as Critical Incident Stress
Debriefing (Mitchell and Everley, 1996). Victims are brought together in groups, and each one is asked to recount the facts of
what each victim remembers happening. Each victim is then encouraged to express his or her thoughts and feelings about the
carnage they have witnessed. Victims discuss any symptoms they
may be experiencing, and the debriefing closes with some suggestions about how to manage the stressful event and what to expect
in the coming days. Such debriefings may last several hours at a
time, and may be repeated as needed.

Critical Incident Stress Debriefing is crisis intervention for whole
communities. It helps the community to normalize the reaction
of various neighbors, provides the beginnings of restored mastery and caring attachments, and provides a sense of purpose as
the community begins to rebuild itself. Critical Incident Stress
Debriefing is a large-scale but very effective intervention for treating
psychological trauma and its aftermath. If you are ever a victim
of violence from large-scale disasters, do not avoid the debriefing
opportunities that can be yours at the time of the incident. If you
have been such a victim and still are distressed by what happened,
contact a local mental health service, and ask them to arrange a
debriefing for you and anyone else who was affected.

Help From the Legal System

Just as families and counselors can be assets for victims in the recovery process, so too can the legal system. This is the system victims can turn to for redress from any of the physical or psychological harms that have been inflicted on them. The legal system provides police assistance and investigation, judicial review, and court trials, when it is necessary. Most of you have some general knowledge of these services, but some of you may not be aware of the legal service specifically designed for the needs of victims. It is called the Victim Assistance Program.

To be as helpful as possible, many legal jurisdictions have begun these Victim Assistance Programs to help victims learn of their rights, and to help victims find the best individual path through the criminal justice system. These programs are funded by fines levied on assailants who are found guilty.

Victim assistants can do many helpful things. They can instruct the victim on how to file an impact statement which lists the financial, physical, and mental health losses that have resulted from the crime. They can put in place any protective services that might be needed, and they see to it that the victim receives any compensation that he or she is entitled to. Assistants are advocates for victims and their families in many useful ways, and are an important resource for any of you who are victims.

Victims, however, must also consider the potential hazards to themselves when legal services are utilized. Many legal proceedings, by their very nature, may serve to re-traumatize victims.

Untreated victims or those who have not fully recovered are prone to flashbacks and intrusive memories of the traumatic event, as we have seen. A detailed, hostile cross-examination in court may biologically re-traumatize the victim. As the victim recounts the episode of violence in specifics, the body's and mind's emergency mobilization systems may again be activated, and may cause the victim to reexperience the terror of the original traumatic event.

A similar process may occur with the media. The news media are intrusive by nature, as reporters doggedly seek to ascertain the facts. Their persistence in questioning the victim outside of court can produce the same harmful effects as the trial itself. Moreover, if the assailant should be found not guilty, the victim who has come forth may experience profound depression for a second time.

Female victims may have an additional general issue to consider. By and large, policing is done by males in a male organization with male values. While the law is moving toward more equality of rights for both genders, it is an imperfect institution and still often favors male assailants. For example, domestic violence is not always treated as a crime that results in arrest (Edwards, 1990).

For these reasons, it is important for you to decide with your attorney, family, counselor, and victim assistant whether utilization of the legal system will prove helpful. Here are some basic questions to consider. Are there medical bills that need to be paid? Is the assailant apt to victimize others? Do you need to go to court to ensure future safety? Has your public reputation been destroyed? Would a trial further humiliate you? Unfortunately, you and other victims still have to weigh the public costs versus the private gains in seeing that justice is rendered.

Specific Legal Issues

We shall now explore some of the traumatic events that are heard by the courts, and some of the benefits and costs in each set of circumstances. If you decide to incorporate the legal system in your recovery, there are certain issues that victims of specific types of crime may possibly encounter. These problems are cited below. They are not exhaustive of all the litigation issues that could arise, and are meant only to illustrate some of the common experiences of victims.

Rape.   Female rape victims may encounter two potential sources of distress: one with the police, the second with the courts.

The first issue involves the female rape victim having to report the rape to a male officer. This is overwhelming for some victims, as they are ashamed of the matters they must recount, and may understandably be angry with all males. This has proven difficult for so many victims that police departments are developing special teams of female officers to deal with rape cases. This is a helpful step forward, but it is not yet available in all cities and towns.

The second problem is with the court trial. Not only will the female victim have to describe the violence in detail, she may also find herself being examined in regard to the cultural myths of rape that we noted in chapter 5. The defense questioning may imply that somehow she asked for it, was behaving in provocative ways, or did not really try to fight off the assailant. All of the personal details of her career, sexual, or financial matters may become the focus of public record and scrutiny.

Each female rape victim must understand and evaluate such matters in her own case. Male rape victims do not usually seek redress through the courts, in large part because of cultural stigma but it is reasonable to assume that they would encounter similar problems.

Spousal Violence. Domestic battering raises equally vexing issues for its victims. Police officers generally do not arrest batterers. They usually walk the assailant around the block to cool off, or may order the assailant to remain out of the home for twenty-four hours. Unfortunately, this approach offers the victim little protection. A police study of domestic violence in Minneapolis, Minnesota (Bouza, 1990) found that making an arrest of the batterer was the most helpful way to stop domestic violence, but police who arrest batterers are still in a minority.

The battered victim may be told to seek a court restraining order. Such orders prohibit the assailant from coming within, say, five hundred yards of the victim. The victim is to call the police if the assailant violates the order. In truth, the police may not always respond to such calls effectively (Edwards, 1990). Rather than be beaten again, women often seek refuge in shelters for battered women. (Some assailants have been known to follow the victim to the shelter, but the police will remove the batterer from the premises.) Male victims of battering usually leave the house till things calm down.

The courts also present difficulties. Female victims will often swear out a complaint against the assailant at the time of the domestic crisis. All too frequently, the woman, who feels frightened, guilty, or reassured that the violence will stop, withdraws her complaint just before the court hearing. Busy police officers who must be in court (often on their own time) become understandably frustrated when such complaints are withdrawn. This practice by victims is widespread, and officers have been forced to prioritize their time and responsibilities.

The perception of female battered victims in the legal system may also lead to other court problems. In the early cases of domestic violence, feminist lawyers and scholars correctly described victims of such domestic violence as the dependent, frightened, non-assertive individuals that we recounted in chapter 6. As medicine has studied the long-term effects of chronic domestic abuse, it has learned that with enough abuse some women will become violent themselves toward the spouse and toward their own children. Our understanding in case law has not as yet kept pace with this understanding in medicine. Consequently, women, severely battered over many years, may find themselves challenged as to why they did not protect their children, and why they have become so violent themselves.

Child Abuse. The abuse of the young demonstrates again how our medical understanding of such events has yet to be incorporated into case proceedings. Here there are several different issues.

Accurate testimony from children is difficult to obtain. They often feel overwhelmed in court. Some states are debating the admissibility of pre-recorded videotaped interviews with the child, but there is as yet no national standard. Neither interviewing children nor using anatomically correct dolls in cases of sexual abuse has proven to be a reliable and valid way of assessing what has actually happened to the child.

A court finding of guilty in any particular case still does not resolve matters, as the judge and jury must then decide what is in the best interests of the child over the longer term of its growth and development. Is it better to remove the parent or to hope that with treatment the abuse will stop and the child's life will be more normal? Society, for example, believes that the mother-child bond is most central to the child's growth. Most courts are reluctant to remove the child from the mother, except when there is evidence of severe maternal substance abuse. In a case of paternal abuse, the court finds itself presented with the same differing opinions of the specialists in this field that we noted for maternal abuse. Determining what is truly in the best interests of the child becomes a very complex matter with no easy solution.

In recent years, the whole issue of alleged child abuse has become more complicated in divorce proceedings, as one parent accuses the other of abusing the child, when both are fighting for custody of the child. The increasing number of instances in which a child is abducted by the parent who did not obtain custody or visiting rights is presenting a new challenge to the legal system, and is complicating our understanding of what does actually constitute child abuse.

PTSD as a Legal Defense in Murder.   We have studied the phenomenon of dissociation in which events later in life can result in flashbacks of earlier traumatic memories. The untreated victim usually appears to have no control over such events, as his or her consciousness alternates between the flashback and present consciousness. If a person commits a crime when he or she is in the flashback state, is that person guilty of the crime?

For example, if a combat veteran hears a helicopter overhead in civilian life and has flashbacks to the war itself, is he or she guilty if he or she shoots an innocent civilian at home because momentarily the veteran thought the citizen was "the enemy"? If a victim had Multiple Personality Disorder in which one personality or altered state committed a crime, would the host personality be responsible for the crime? Such cases are now before the courts, but there are presently no clear standards to help us decide culpability in such matters. Each case is decided on its own merits as society, through its courts, struggles to integrate recent findings in human behavior in a way that is just to both victim and assailant.

Spousal Murder.   Murder is usually defined as the premeditated (thought out beforehand) use of excessive force (more than is needed for self-defense) when there is no imminent (immediate) harm. We have noted in chapter 6 that female victims of severe battering usually kill the assailant after thinking about it, and when there is no imminent harm (because he is resting or asleep). A gun or a knife is often used, and in these circumstances is considered to be excessive force because there is no immediate life-threatening event that requires self-defense.

Are these women killing in self-defense, if the victim is at rest? These are legal issues the courts are currently attempting to decide. Until recently spousal victims were usually found guilty of premeditated murder. In more recent times, governors in various states have commuted the sentences of such victims as society begins to realize that some victims of long-term battering are confused and fearful of being killed. In such a state, spousal murder may be the only alternative for self-defense that they comprehend. The health sciences will need to spell out clearly in the coming years the consequences of long-term abuse on reasoning, memory, and motivation so that justice will be ensured for everyone.

*   *   *

Each of you should consider the issues involved in utilizing the judicial system. This is not to suggest that you not use the legal system but only that you make informed choices when you do. The courts can be helpful, but in many cases the rights of the assailant still take precedence over the victims rights as the courts seek above all to protect the innocent. Victims do not always receive justice, and you should consider this possible outcome.

Prevention

We all need to be mindful that we live in an era of increasing violence, and that we should take whatever steps we can to protect ourselves from being victimized or re-victimized. Anthony Bouza (1990), former Chief of Police in Minneapolis, Minnesota, has written about these matters and has some suggestions for us to consider. I include them here.

Chance is a basic consideration in criminal behavior, as assailants engage in such activity impulsively, when they think they can get away with something. A second factor that Chief Bouza has noted is the fact that many of us do not often know when we are in geographical areas of high risk. Without this awareness we may unknowingly venture into high-crime neighborhoods, into darkened alleys, and into situations with other people whom we should be avoiding. A third factor contributing to becoming a victim is that many people are high risk takers. They feel they can continually beat the odds, or that nothing will happen just this one time, and so they engage in activity that puts them at high risk. A final grouping of risk factors relates to certain personality characteristics that place the person at increased risk. Believing the best of people, being open with others, and easily trusting others can place us in situations where we can be taken in by criminals and assailants of all types.

What should we do to protect ourselves? Here are some useful suggestions (Bouza, 1990; Warsaw, 1988):

1. Stay sober. Much crime and many traumatic events occur when the victim's senses, memory, concentration, and general reasoning have been dulled or anesthetized by alcohol or drugs. A common example that we have discussed is rape. Many rapes occur when the victim and the assailant are both drunk. Avoid victimization by remaining sober. If you are using alcohol or drugs, get into a recovery program as soon as possible. You are at higher risk while you are addicted to substances.

2. Pay attention to the environment. Know the general safety of the places where you go, and learn of any risks as you visit new places. Pay attention to the people about you, and protect your valuables and your person in your home and in your motor vehicle. Pay close attention to your intuition. If you feel unsafe or ill-at-ease, your body is responding to some potential threat that you may not be fully conscious of as yet. Remove yourself from such circumstances, go to an environment where there are other people, and be prepared to call for help. Be conservative in such matters. It is better to call for help and learn that you were incorrect, than not to call for help and learn that you are a victim.

3. Stay away from isolated places. People who are by themselves in elevators, corridors, stairwells, abandoned streets, and so forth increase their risk of being preyed upon. Decrease your risk of becoming a victim by avoiding such situations. Keep your eyes and ears open in those instances when you must be alone.

4. Avoid high risk taking. This is especially true for thrill seekers, but it is also the proper mind-set for all of us. With the crime statistics going up, any of us makes a statistically losing bet when we feel we can get away with the risk just this once. More and more of us are losing the gamble. Better to learn self-defense, to be with others, and to enjoy the many things in life that involve no serious risk.

*　　*　　*

We have now reviewed the important individual steps you can incorporate into your recovery program as well as the three support networks that can provide additional support in restoring reasonable mastery and caring attachments.

We have also spent some time exploring the first steps in restoring a sense of meaning and purpose in life. We have seen the importance of reconstructing what actually happened, trying to understand your response to the event, learning how to grieve its loss, accepting whatever new limitations it may have imposed on you, and reflecting on your basic values.

In the final chapter we shall complete the remaining steps in the process of finding renewed meaning in life. We shall examine the nature of violence and then discuss the importance of loving others as a way of moving on in life. The stress-resistant characteristic of being concerned with the welfare of others provides an important avenue for putting the anger, depression, and bitterness of violence behind you.

10

Recovery: Why Me?
Searching for the
Soul of Goodness in
Things Evil

*Never shall I forget those moments that murdered my God
And my soul, and turned my dreams to dust.*
— Elie Wiesel

There is some soul of goodness in things evil.
— William Shakespeare

The bells tolled softly on this fog-shrouded morning in the city by the sea. The calls of the gulls were a chorus of lament as the friends and relatives of eighteen-year-old Darryl Redford Junior climbed the damp steps of the First Baptist Church, and followed his forever stilled body down the center aisle. The media attention as the mayor arrived only added to the agony of painful defeat.

As the minister spoke, Emma Redford thought back eighteen years before to the delivery room. She had felt so alone then and feared that she would die in childbirth. She knew that such things happened more frequently to minorities and the poor, and that she was a part of each of those worlds. Emma had named her son Darryl after his father, the father who had abandoned them after learning of the pregnancy.

Her own life had not been easy. Sexually abused by her uncle when she was young, never having known her own father, Emma and her mother had fought their daily struggle to keep body and soul together. And struggle it was. Poverty, ill-health, continuing racial slurs. Problems were everywhere, but their belief in education and in God had kept them going.

Darryl Senior had been the fulfillment of her youthful dreams and prayers. How she had enjoyed those newly married hours that they had spent planning their future together. Her body longed to embrace his soul. A new beginning. A chance to erase past hurts. Her soul was crushed when he withdrew.

She turned to God and to her new baby. She put aside her own dreams for those of Darryl Junior. His life would be better. And so it was. A decent kid, he was active in sports and diligent in his studies. Words could not express her joy when in his eighteenth year he had won a full scholarship to the state university.

She had sent Darryl and his best friend Pogie to the convenience store for milk. It was there that their young lives ended. One bullet to each head. Innocent victims of a botched robbery attempt that netted the teenage thief seven dollars and seventy-seven cents.

"Why?" asked Emma in the silence of her heart. "Why me?"

The fog remained damp and thick.

*   *   *

In her grief Emma is asking a question that most of us have asked ourselves at some point or other in our lives. Why? Why has this befallen me?

The cry, "Why me?," is a plea for meaning. The victim is attempting to make some meaningful sense of the traumatic event that made no sense when it happened and often makes no sense even after the victim has struggled with it for years. The victim, the victim's loved ones and friends, and the victim's counselor have come face-to-face with the mystery of evil. Despite the fact that mysteries are situations that we can never fully understand, the effort to find some meaning in the event helps us to regain some sense of mastery and to find a new or renewed purpose in life. Even those victims who lack a belief in the transcendent often have a view of the world that can help them make some sense of what is most often irrational. One way or another, all of us who become victims must work to clarify what these painful events mean for our lives.

In this chapter, we shall take a few hesitant steps toward a possible answer to the question, "Why me?" First, we shall explore the mystery of evil in an attempt to gain some understanding of the pain and suffering associated with natural disasters and with the malicious acts that some humans inflict on others. Secondly, we shall focus our attention on the nature of love, what loving others means, and how loving others can heal the pain and suffering of traumatic events, and provide you with a fresh, meaningful purpose in life. Loving others can aid you in your search for the soul of goodness in things evil.

Some Thoughts on Evil

Evil consists of the seemingly senseless acts of destruction that occur in our lives at the hands of nature or at the hands of others.

Much of the trauma in the world is a result of natural catastrophes: floods, volcanoes, forest fires, tidal waves, earthquakes, hurricanes, and drought leading to famine. These are part of the nature of things, givens that cannot be changed, and we have to accept them as facts of life and take what precautions we can. In fact, much of the trauma-related research suggests that, while victims of natural disasters are angered, they are equally awed by the power of nature. In time, such victims seem to recover somewhat more easily than victims of human maliciousness. It is easier to be humbled by nature than to be personally humiliated and degraded by another person.

For most victims abused by other persons, such human viciousness and cruelty cannot be easily explained away as givens. In cases of sexual abuse, battery, murder and so forth, we do not seem to be facing the randomness of natural catastrophe, but rather the deliberate acts of twisted human beings. In the face of malice, madness, and perversion, the search for meaning becomes more difficult. In understanding human cruelty, we need to begin by drawing a distinction between human violence that is the result of true medical illness, and human violence that seems to be willfully inflicted.

Medical Factors in Human Violence

Some common medical conditions may contribute to violent acts. Some preclude human responsibility and volitional choice; others do not. Knowing that a particular act of violence by another was the result of a medical illness can sometimes make it easier for the victim to bear. Police, counselors, lawyers, family members of the assailant are sometimes able to provide victims with this information. In case you have such knowledge about your assailant, I have listed below some of the more common medical problems that may lead to violence.

1. Temporal Lobe Epilepsy. The temporal lobe is that part of the brain that is involved in memory storage and retrieval. Disruptions in brain chemistry or electrical conductance in this part of the cortex will result in seizures. Seizure behavior in some persons is associated with aggressive outbursts and assaults on others. Such aggression is not under the control of the individual.

2. Major Mental Illness. There are some medical diseases that affect the mind and brain chemistry. Schizophrenia and bipolar disorder are two common examples. These are genetic illnesses that run in families. Under stress, changes in brain chemistry may interfere with normal reasoning processes, and can lead to hallucinations which are altered perceptions that are not true, such as the hearing of "voices" outside the head, and to delusions, which are false beliefs about the self. The voices often command the individual to harm himself or herself or others, and the delusions that are often persecutory in nature will lead the person to strike out against a perceived enemy.

When such individuals break the law, if their major mental illness was actively present at the time of the crime, the individual is often found not guilty by reason of insanity. The schizophrenic or bipolar patient has no more control over these flare-ups than an arthritic patient can control the flare-ups of that disease, and the insanity plea is society's recognition that the medical illness was beyond the control of the patient. There are some medicines that can help these illnesses, but the medicines are not always foolproof. They are of little help in cases where the patient is delusional, and refuses the medicine because the patient believes it will poison him or her.

3. Alcohol and Substance Use.   Alcohol use is very often present in violent behavior. Eighty-five percent of all suicide attempts involve alcohol abuse, and much domestic battering is done by drunken spouses. Enough alcohol will disinhibit the brain's control systems and lead to destructive behavior. In addition, some alcohol users have a medical condition known as pathological intoxication. Their brain chemistry, for reasons that are not as yet clear, react almost immediately and negatively to the presence of alcohol in their blood. The brain's control systems appear more easily disinhibited in these persons, and violence, usually towards others, is often predictable.

Certain street drugs have been found to have similar disinhibiting properties, and are likewise associated with increased violence. Medical research has demonstrated this possibility of loss of control in the use of amphetamines, barbituates, cocaine and crack, and PCP or phencyclidine. Substance use can lead to violence and the user is usually held accountable.

4. Posttraumatic Stress Disorder.   Paradoxically, PTSD itself can lead to violent outbursts in victims of abuse. In a recent study of over one thousand incarcerated male prisoners with no past history of combat experiences, those prisoners with PTSD symptoms had a seven times greater chance of having been arrested for a violent offense within the preceeding twelve months than those without such symptoms. In eighty-five percent of these cases, the individual experienced his symptoms of PTSD just prior to or during the violent behavior. Similarly, there are many documented examples of adult males with a history of abuse who are sexual predators of young males. These adults themselves have a history of being sexually abused in their own childhoods.

It is not clear why this should be so. Perhaps some of these individuals encounter a stimulus in the environment that leads to flashbacks and dissociative experiences. Others may be coping with the kindling phenomena or with endorphin opiate-like withdrawal. The role of personal responsibility in such situations must be addressed on a case-by-case basis.

5. *Organic Personality Disorder.*   Some individuals exhibit marked increases in hostility and aggression, impaired social judgment, apathy, and suspiciousness after biological trauma, strokes, or head injury. These outbursts can become daily patterns of behavior, and several causes of this personality disorder have been identified. The causes may include brain tumors, viral infections such as encephalitis or rabies, cerebral vascular accidents, seizures, Multiple Sclerosis, Huntington's Chorea, some forms of mental retardation, and substance use. The side effects of some prescribed medicines may also produce organic personality syndrome.

If the patient is under twenty years of age, seizure or head trauma are the more likely explanations; if over fifty years of age, cerebral vascular accident, Alzheimer's disease, or chronic alcohol use are common causes. Patients with Organic Personality Disorder are not usually considered responsible for the violence.

6. *Intermittent Explosive Disorder.*   This disorder is characterized by sudden, unexpected outbursts of aggression out of proportion to the event that appears to induce the anger. Such episodes pass quickly, and the patient becomes calm and remorseful. These individuals are not intoxicated, are not suffering from a major mental illness, and are not antisocial. The cause of this disorder is not fully understood. There appears to be some dysfunction of the limbic system or possibly some form of psychomotor epilepsy. Individuals with this disorder are again not usually considered responsible for these aggressive outbursts.

7. *Antisocial Personality Disorder.*   This disorder involves a long-term pattern of aggressive behavior toward others that we spoke of in chapter 4. These individuals disregard society's rules and values.

The research findings to date about antisocial individuals do report abnormal electroencephalogram (EEG) or brain-wave readings, and a higher incidence of problems paying attention to tasks at hand (Attention Deficit Hyperactivity Disorder). Twins studies have found a higher than normal rate of criminal behavior among blood relatives of sociopaths.

These biological findings do not excuse sociopathic behavior, however. Since there are many individuals with abnormal EEGs, attention deficit disorders, and so forth, who are honest, well-intentioned men and women, it is reasonable to assume that antisocial behavior may be a function not only of biology, but also of willful intent.

9. Conduct Disorders.   Conduct disorders are diagnoses made on children under the age of fifteen. Sadly, we are learning that prepubescent children are more and more involved in aggressive behavior, including the four major adult crimes of rape, assault, burglary, and murder. It is reasonable to assume that such trends will increase.

As you might expect, all of these children have difficulties at home. Home-life is marked by an absence of caring attachments to family, other adults, peers, and friends. These children have temper outbursts, mood irritability, and problems with substance abuse. Some have an Attention Deficit Hyperactivity Disorder. Grades in school are weak, and all of them have poor self-esteem. Often they adopt a stance of "toughness" to compensate for their feelings of inferiority and mistrust.

Why these children behave in this way is not known. Temper outbursts, mood irritability, substance abuse, and withdrawal from others suggests that some of them may be suffering from PTSD, and there are cases of conduct disordered children who have in fact been victims of rape, incest, or assault. In other cases, faulty parenting, and modeling the aggressive behavior seen at home may be as important to the outcome of conduct disorder as any biological risk factor.

Conclusions.   Clearly there are some biological and medical conditions that place the afflicted individual at risk for violence, and diminish that person's responsibility for such acts, but this is not true for all biological conditions. Table 1 lists the basic biological problems associated with aggression and violence, the usual age of onset, and society's judgment about whether the person is responsible for his or her actions or not.

As we can see from table 1, temporal lobe epilepsy, major mental illness, organic personality syndrome, and intermittent explosive disorder are usually considered to be beyond the person's control. PTSD-related violence must be judged on a case-by-case

Table 1

Medical Factors in Human Violence:

Medical Condition	Age of Onset	Responsible for Actions
Temporal Lobe Epilepsy	Any Age	No
Major Mental Illness	Late Teens Onward	No
Alcohol/Substance Abuse	Any Age	Yes
Post-Traumatic Stress Disorder	Any Age	Case-By-Case
Organic Personality Syndrome	Under Thirty/Over Fifty	No
Intermittent Explosive Disorder	Twenties or Thirties	No
Antisocial Personality Disorder	After Age Fifteen	Yes
Conduct Disorder	Before Age Fifteen	Case-By-Case

basis because of the possibility of dissociated states. No one has to be drunk, or has to abuse substances. In acts committed in intoxicated states, the individual is usually held accountable because he or she ingested the toxic substance by choice in the first place. Adults who commit antisocial acts are held accountable, and the courts currently are deliberating how responsible young children are, and at what age, and for which crimes they should be considered adults.

How are you, the victim, to understand the evil that comes from these medical illnesses? It will be helpful to you if you think of these acts as similar in nature to natural disaster. Like natural disasters, these types of violence are part of the nature of things, part of the givens we cannot change, and, like victims of natural disasters, you were the wrong person in the wrong place at the wrong time. There is no fault or evil in you. To the contrary, we often need to put away such people to protect them as well as others.

Psychological Factors in Human Violence

In addition to the biological factors that can lead to aggression, there are many psychological conditions that may also result in violence. The psychological reasons for aggressive behavior most often appear to be deliberate, willful acts of malice to inflict harm on others, and do *not* appear based in medical illnesses. While the reasons for the violence are varied, once again, there is no fault in the victim. When you spot these types of persons, get away from them as soon as possible.

Some individuals' belief systems lead them to violence. Proponents of some political, social, economic, and religious causes believe they have an absolute right to impose those views on others by force, if necessary. In other examples, the reasons are more directly personal. Some abuse power to exact revenge for real or imagined injustices. Having been hurt once, they are determined never to let it happen again. Others fear being dependent on other persons for fear of losing their sense of self. Such persons at times use aggressive behavior to keep others at bay. Some trust no one, or have excessive fears of being abandoned or rejected, and similarly use violence to drive others away. The reasoning is usually as follows: I will leave you before you can leave me. Others feel guilty, but find it so painful to accept in themselves that they project the guilty feelings onto others, and then lash out at them as punishment. Some besieged individuals engage others through violence to maintain attachments when more constructive methods of interacting have failed. Still others engage in random acts of violence because they are bored, angry, or without meaningful goals or tasks in their lives.

Finally, many perpetrators of violence are driven by jealousy, envy, sadism, and prejudice. These are strong negative human passions, and reflect a maladaptive and overpowering need to control others and the world at large. Some of our most violent, evil acts of human aggression stem from these passions, and we will spend a few moments examining them.

Sadism/Mental Cruelty.   Sadism is the infliction of physical pain, psychological hurt, and/or humiliation, and mental cruelty is the deliberate attempt to humiliate or hurt another person's feelings. People who do these things are afraid of life because they cannot control it. Their excessive need for control is rooted in a sense of complete incompetence. Sadism or mental cruelty temporarily transforms this sense of ineptitude into power and the illusion of control. Such control steals another person's freedom (Fromm, 1973), but it appears to temporarily reduce the perpetrator's sense of panic.

Jealousy/Envy.   People feel jealous when they fear that another more powerful person will take from them what they feel helpless to protect, for example, feeling jealous that someone will steal your boyfriend or girlfriend. People feel envious when they are angry at what someone else has but they themselves feel powerless to attain. For example, some people feel envious of the success of others. Jealous and envious people often seek revenge in the form of slander and gossip to ruin another person's reputation. Both jealousy and envy in adults appear rooted in childhood problems where the envious or jealous person felt another sibling was more favored by the parents. This led the envious or jealous person to feel abandoned by the parents. In adult life this anger toward the favored sibling is directed to some other innocent adult who is considered to be favored in some special way. Jealous or envious individuals seek to ruin the other's reputation to make everyone equal and to have the power to prevent themselves from being rejected yet again.

Hatred/Prejudice.   Hatred is intense hostility or antipathy toward a person or group of persons. People hate when they have no goodness, justice, or love in their lives. To be sure, sometimes hatred is rational and not evil as when there is a true objective threat, but more often hatred arises after a long series of bitter disappointments in the hater's life. Hatred is derived from the Greek word for grief, and often these individuals have been rejected by a parent, have poor self-esteem, have economic insecurity, or are not able to attain the sexual partner of their choice (Allport, 1958).

Prejudice is the placing of these hateful feelings onto another person who is somehow different. Acts of prejudice can include

slander, avoidance, discrimination, physical attack or extermina-
tion, and often the prejudice is directed toward whole groups of
individuals *qua* groups. Race, color, creed, gender, and social class
are common factors on which prejudicial decisions by individuals
are based.

Group Evil.   Humans are social beings, and form groups natu-
rally based on comfort and similar interests. Just as prejudiced
individuals can commit evil acts, groups of humans can behave in
a similar manner, if they too lack a basic sense of justice or love
in their lives. They can develop a sense of invulnerability and
moral superiority, and this arrogance can lead to willfully chosen
acts of destruction toward other innocent persons. The group di-
rects its fears and prejudices onto the "enemy." Since the group
is usually removed from the enemy, the enemy is easily dehuman-
ized, and evil acts follow. Since the group is a closed entity, there
is no feedback loop to correct these fundamental misperceptions
that they have of others (Sanford and Comstock, 1971), and the
prejudice and hatred continue.

The Mystery of Evil

How then are we to make sense of twisted human acts of destructiveness? As we have noted, those acts of aggression that are
based in medical illnesses can be understood for what they are:
aberrations in normal body chemistry that result in random acts
of violence. The psychological acts of violence are harder to understand. It would be easy to explain away the adult anger as a
function of an unhappy childhood, but many persons have equally
unhappy childhoods and do not commit such acts of senseless
destruction. How are we to understand such violence? Society has
some different ways to understand the mystery of evil.

Some find it helpful to believe in fate, and to accept that the
principles of good and evil are givens that we must accept. Others
believe that there are basic rights for individuals or basic ethical
values that require us to treat each other with basic justice, love,
and respect. Still others believe in a loving God who has given
humans freedom of choice that can be abused when individuals
act in ways that are destructive to others and themselves.

Each of these points of view can help us find where evil might fit into our understanding of a meaningful world. These perspectives can help us balance the horrifying experience of malice in the world with an awareness of the goodness that exists in many places in that same world. These viewpoints, of course, will not fully answer the question of "Why me?" nor fully explain the mystery of evil, which remains a dark muddle of the irrational in human beings.

In a way there is no answer to the question "Why me?" Yet recognition of abuse and the struggle for integrity bring many to the way of loving others. Many victims can tell us of how horrible incidents in their lives ultimately opened up for them worlds of new possibilities, and openness to others. Loving others heals traumatic wounds, and can provide a transcendence in life that restores in you, the victim, a purposeful meaning in life. I have seen this happen many times in my years of counseling. As Elisabeth Kübler-Ross has written, to love means never having to be afraid of the windstorms of life.

Some Thoughts on Love

What does it mean to love another person? Love may be defined as helping another person grow to his or her potential without expecting anything in return as long as neither party is being physically or psychologically abused. Loving others is the reverse of evil. Rather than consciously seeking to destroy the other, the one who loves is working with the principles of growth in the world to foster the development of another person. In this definition, love is an act of will, just as evil is often an act of will. Loving others helps us keep physical and moral evil in perspective. Loving others allows healing to occur. We become whole and renewed again. Often our love for others leads us to a deeper understanding of God. Love, with its concerns for others, is experienced as a sense of contentment or inner peace.

There are three parts to this definition of love. The first is helping the other person grow to his or her potential. The one who would love needs to consider the assets and liabilities of the other and to determine how best to foster that person's growth. Sometimes this is done by teaching some form of mastery, at other times by listening quietly as the other seeks to find himself or herself. At still other times it may mean setting limits as the other ventures down a path that is not in his or her own best interest. In each step in this process, the one who would love continues with his or her own life and is careful to act as a catalyst in helping the other seek his or her goals. Such efforts require constant care and attention.

The second component requires that the one who would love expects nothing in return. In an age of self-gratification this may be difficult, but it is essential for the other person's growth. The one who loves is there only to facilitate and enable. We cannot expect the other to help us in difficult times, or to reciprocate our helpful gestures, or even expect that we should feel good about helping. Our efforts are freely given to the other with no strings attached, no reservations. It would be pleasant if such reciprocity occurred but often it does not. We can expect nothing in return.

The final component concerns the physical and psychological well-being of all concerned. A relationship in which *either* party is being physically abused, verbally denigrated, or is severely neglecting his or her own growth and needs is not a caring, loving relationship. One must love one's neighbor as one's self which means that the person who would love must also not neglect his or her own personal growth. Abusive relationships do not meet the basic standards for safety, health, and well-being, and such relationships should not be allowed to continue. In such cases, discontinuing the relationship may be the most caring thing to do, as the presence of the one who would love permits the other to act in evil ways and to destroy the very growth the one who would love seeks to foster. True love results in increased growth and self-esteem, not the reverse.

Being concerned for the welfare of others is an abstract concept, but this love for others becomes more meaningful in our specific interactions with particular individuals. In seeking out specific persons, we make the abstract real, and we avoid the loneliness that comes from being alienated and cut off from others. Loving in these concrete ways is especially important for victims, since traumatic events by definition may cut us off from caring attachments and the purposeful meaning in life that such attachments offer.

Searching for the Soul of Goodness

With a deeper understanding of violence and evil, each victim is now ready to complete the process by finding a transcendent meaning in life that provides meaning for all that the victim of PTSD has been through. Loving others in specific concrete ways, helping other people to grow as a result of your suffering, is considered to be one of the better ways of finding a new purposeful meaning. It is the pathway to becoming stress-resistant, and here is how you can put this process to work for others and for yourself.

Just as each victim experiences evil, pain, and suffering in real and abstract terms, so should each victim strive to love others in concrete circumstances. Whom would you like to love? Whose growth do you wish to foster? Where will your efforts be received? Of all of the people you have encountered in your life or that you may know of but have never met, who needs you? Someone really does, and victims should direct their energies and efforts toward those who are in need. There are many ways to love others. Specific acts of kindness, educational and political efforts on behalf of victims, becoming a health care provider are some of the many ways transcendent meaning may be found. These ways are as healing for victims as for those who will be loved, and it enables victims to find a renewed sense of purpose in living.

Patients, students, family, and friends have persistently asked me over the years to be more specific and concrete about how to actually love others. My suggestions to be helpful where you can, be kind, be thoughtful of another's needs were often considered to be too general in nature. It was one of my patients who reminded many of us of the specific ways to love for which so many search.

Mildred died in her mid-seventies. It was a slow and painful death from Alzheimer's Disease. Mildred had first come to see me ten years earlier when her husband of almost forty years had died. She was understandably quite depressed. She herself had been a victim of early childhood abuse at home, and had risen above it by finding meaning in life in her marriage. Mildred grieved her painful loss, and began to devote her energies to homeless children.

A few years later she returned to see me again. This time she had memory problems. We were both to learn that this was the beginning of the Alzheimer's Disease, whose complications would take her life several years later. As is the case in Alzheimer's Disease, her cognitive capacities for thinking, reasoning, and communicating gradually failed. She was a prisoner within her own body. It was as painful for her to realize as it was for all of us who knew her.

Her funeral service was sparsely attended because she had outlived so many of her friends. Her niece, the staff, some of the patients where she had spent her final years, and the minister were present.

Unknown to all of us, Mildred had planned her own funeral service before she had begun her descent into silence. She had chosen her favorite passages from the Bible to leave us questions about life to ponder. She asked very basic questions of herself and of us: What is the meaning of life? Who is God? What is the correct path for a human to follow? And, not surprisingly, What does it mean to love another person?

Here is Mildred's choice about specific ways to love others that she wanted us to remember.

If I speak in the tongues of men and of angels, but have not love, I am a noisy gong or a clanging cymbal.

²And if I have prophetic powers, and understand all mysteries and all knowledge, and if I have all faith, so as to remove mountains, but have not love, I am nothing.

³If I give away all I have, and if I deliver my body to be burned, but have not love, I gain nothing.

⁴Love is patient and kind; love is not jealous or boastful;

⁵it is not arrogant or rude. Love does not insist on its own way; it is not irritable or resentful;

⁶it does not rejoice at wrong, but rejoices in the right.

⁷Love bears all things, believes all things, hopes all things, endures all things.

⁸Love never ends; as for prophecies, they will pass away; as for tongues, they will cease; as for knowledge, it will pass away.

¹³So faith, hope, love abide, these three; but the greatest of these is love.

To be a survivor of violence and abuse is a noble human goal, and the reason for which this book was written. What happened to you was not your fault. If you follow the steps that we have outlined and implement them with the help of others, you will in time be safe, the ugliness will not happen again, and the painful memories will diminish.

My hope for each of you, each victim, is that in loving others you will find the soul of goodness in things evil, and that your dark and lonely night will come to an end.

APPENDIX A

Your Project SMART Program

Over the years, my patients, my students, and registrants in workshops I have offered for business and professional groups have asked me to find some way for them to get started with all these possible strategies for developing stress-resistance, and to help them focus on a manageable place to begin. In response to these requests, I have developed Project SMART (*Stress Management And Relaxation Training*):

Project SMART'S goals are basic: (1) to reduce life stress by deactivating the stress response; (2) to teach some strategies for reasonable mastery; and (3) if you do this in a group with others, to develop caring attachments. Although you could do Project SMART alone, it is more effective and more fun if it is done with others.

Project SMART includes gradual reductions in the dietary stimulants (caffeine, nicotine), relaxation exercises, aerobic exercise, and an approach called stress inoculation in which you can rehearse beforehand better strategies for solving problems.

To begin this program, gather some of your friends together. Each of you needs to have a physical exam to make sure you have no medical problems, and to be cleared for the aerobic exercise component. Group meetings take about one-and-a-half hours each, and you need eight to ten meetings to experience real benefits.

The first meeting is devoted to a general discussion of life stress in our current age. Avoid personal psychological problems and traumatic episodes; rather, discuss the challenges of daily life facing all of us: incompetent drivers, limited day care, finding a reliable repair person, and the like. The remainder of the first session focuses on the process of how life stress occurs, and a presentation of the four stress management strategies that I have just outlined and that will be used in subsequent group sessions.

The second meeting begins with a general review of managing stress. The group members each choose a dietary stimulant to cut back on; at subsequent meetings they will review their progress in reducing their intake of the stimulant. If one smokes twenty cigarettes a day, he or she should decide to cut back to eighteen. If another member usually has six cups of coffee, he or she should commit to make one of them decaffeinated, and so forth. The changes should be in small, gradual steps to ensure some mastery and to avoid withdrawal symptoms. Caffeine cutbacks should be about twenty-five milligrams of caffeine (generally one cup of coffee or equivalent). The second session closes with the relaxation exercises found in appendix C. You and your friends should sit comfortably in a quiet room where you will be undisturbed for fifteen minutes. In my groups, we measure changes in levels of tension with a very inexpensive biofeedback skin thermometer known as a Biodot. The Biodot changes colors like a mood ring as the person wearing it becomes more relaxed. Members should use the relaxation exercises at least once a day. Some members may initially experience temporary light-headedness, tingling, numbness, or warmth in their bodies. These temporary feelings are signs that the body has been very tense, and is beginning to relax. If you feel out of control, or experience intrusive, unpleasant thoughts when you attempt to relax, do not do these exercises. Aerobic exercise, which we shall discuss next, is an equally helpful way to reduce stress, and you will feel more in control.

During the third meeting, members discuss their success in decreasing the intake of whichever dietary stimulant they are focusing on, and they agree to continue. The group then does fifteen minutes of relaxation. The remainder of the time is devoted to selecting some aerobic exercise to be done before the next session. Choose something you would like to do (consistent with any directions from your physician), and begin in small gradual steps. Always do warm-up and cool-down exercises. Instructions for aerobic exercising may be found on p. 205. Walking is a good form of exercise to start with, if nothing else is appealing. *Stop if you feel faint, are short of breath, or have pain.* See your physician if you have any of these problems. The aerobic exercise goal is three twenty-minute sessions over the course of seven days. When each member has made his or her choice for the coming week, the group should warm up, go for a brisk fifteen-minute walk, and then cool down before the session ends. Members do the exercise they have chosen between sessions.

If you are a jogger, it is better to run before 8:00 A.M. or after 6:00 P.M. during the summer to avoid air pollution, especially ozone. If you have panic disorder, do not do aerobic exercises without consulting your physician.

Meetings four through ten follow this basic format: discussion of members' progress in implementing small and increasing reductions of dietary stimulants, relaxation exercises for fifteen minutes, followed by a brisk fifteen-minute walk. (Members are expected to be reducing their intake of dietary stimulants, and performing relaxation and aerobic exercises between sessions.) The remaining time in these sessions is used for stress inoculation.

Stress inoculation is a process that trains individuals in basic strategies for coping with aspects of life stress. Sometimes these strategies include how to be a better listener or how to communicate your affection or concern for another. Often, however, these strategies help us in dealing with noncooperation. Noncooperation is a potentially stressful interaction between two or more people in which at least one person does not care about reaching a reasonable solution to the source of the conflict. It is a normal part of everyday life that we must face and with which we must learn to cope. Practicing how to deal with noncooperation in your Project SMART group can be instructive for victims.

Group members should pick a common stressful problem in living in today's age (again, not highly personal problems). Such problems might include dealing with rush hour traffic, communicating with angry store or bank clerks, or tactfully asserting yourself when individuals cut into a line of people waiting to buy movie tickets.

The group members then spend time sharing the strategies they would use to solve the problem at hand. Emphasis should be placed on which solution is the best one for the specific person in a specific context. The group needs to pay attention to what verbal messages the person will give, what really needs to be said to get the problem solved, and how this can be communicated tactfully. Members also need to pay attention to the nonverbal communication of the person who is attempting to solve the problem. If you are returning faulty merchandise to a store clerk, in addition to clearly asking for your money back, you need to be able to look the clerk in the eye, stand erect, shoulders back, and speak in a reasonable and audible tone of voice.

Once the verbal communication is clear, and the nonverbal communication is appropriate to the situation, there is one step left. This step is so important that I insist that my group members not practice their answer to the life stress problem on their own until we have completed the last part as a group. In the last part, the members rehearse the problem with all the possible responses the person may encounter when he or she attempts to solve the problem. The possible responses range from cooperation in solving the problem, to cooperation but with anger, to indifference, to hostility, to complete refusal to cooperate. If the problem is how to deal with the store clerk over faulty goods, some member should role-play the store clerk; another, the customer. They would then practice the varying degrees of cooperation. The remaining group members should offer suggestions for dealing with the various types of noncooperation. Stress inoculation works in a similar manner for situations of expressing caring.

Project SMART has proven itself helpful in reducing stress overdrive, and in enhancing mastery and attachment. It can be helpful to you also. You can improve your stress resistance by utilizing Project SMART in small gradual steps. If a reasonable trial of Project SMART produces no real beneficial effects, see your physician for further professional help.

—Flannery, R. B., Jr. *Becoming Stress-Resistant Through the Project SMART Program.* Ellicott City, MD: Chevron Publishing Corporation, 2003, pp. 115-118, 199-200.

General Instructions for Aerobic Exercise

(1) See your physician to obtain medical clearance before you begin an aerobic exercise program, and always comply with any suggestions or limitations that your physician may give you.

(2) Your aerobic exercises should begin with a three-to-five-minute warm-up period. Slowly walk, or jog or run in place. Bend, stretch, twist, and generally limber up your body. The warm-up period loosens muscles and joints, increases circulation, and helps to prevent injury. Stretch passively your major muscle groups. For those with greater interest, specific diagrammed warm-up stretch exercises may be found in the two references listed below.

(3) Now begin your aerobic exercise. Be sure that you have chosen an exercise that is fun for you to do so that you will be motivated to continue. Start in small, gradual, and manageable steps. *Stop if you feel faint, have pain, or experience shortness of breath.* A healthful exercise goal to work toward is three twenty-minute periods of such exercise on three different days in any one calendar week.

(4) Finish with a cool-down period when you have completed your aerobics. The cool-down period is similar to the warm-up period. Again, for three to five minutes walk about slowly, or jog or run in place slowly. Also do some mild stretching for specific muscle groups. The cool-down period allows your body to adjust from intense exercise to its more normal resting pace.

References

Greenberg, J. S. *Comprehensive Stress Management.* 2nd ed. Dubuque, IA: William Brown Publishers, 1986, 206-207.

Johnson, S. B. Walking Handbook. Dallas, TX: Institute for Aerobics Research, 1989, 26-27.

APPENDIX B

Relaxation Exercises for Victims

The relaxation exercises that I have outlined in detail below have proven very helpful to a great many men and women seeking to reduce the unwanted distress associated with the aftermath of psychological trauma. These exercises can be helpful in reducing the physical symptoms of PTSD, and in minimizing the negative effects of kindling and the endorphin opiate-like withdrawal symptoms.

The complete exercises outlined here are not for all victims, however. Some victims feel out of control when they are relaxed. For others, the very words "trauma" or "victim" cause these abused individuals to tighten their muscles. For still others, the relaxation components of holding one's breath or closing one's eyes may remind some victims of the crime they experienced.

If you feel any of these things may be applicable in your case, there are other ways to relax. My colleague, Dr. Mary Harvey, and I have found that listening to soft music, drawing, knitting, even sitting quietly can be equally helpful alternatives in creating a sense of relative calm. Some victims may wish to use aerobic exercises. (Aerobics are not helpful for victims with panic disorder because they will increase the frequency of the panic attacks.) Many victims may find deep breathing, prayer, or repeating positive self-affirmations to be helpful.

Here than are the complete basic exercises. Modify them if you need to.

*   *   *

The basic relaxation exercises are helpful because they lower the emergency mobilization response all at once. In addition, you can use these exercises in public or private without anyone knowing. These exercises are portable and you can take them with you, and best of all there is no cost to purchase them other than your own time in learning to use them. If you use these exercises for as little as ten or fifteen minutes a day, you will feel remarkably better in a short period of time.

The exercises that I will present here are an amalgamation of deep breathing, release of muscle tension, and the use of pleasant imagery. I have taught these exercises to thousands of people over the years, and this appears to be the best combination for the greatest number of people. As we have noted, there are several different types of relaxation exercises available, and you may want to add in some other types.

In doing these exercises, you will be safe and in control. If a true emergency arose, your mind and body would immediately rise from the relaxation state, and you would be capable of addressing the problem. If you have lung disease, check with your physician before you begin these exercises.

As I have noted, this relaxation program contains three parts: slow-paced breathing, the cognitive release of muscle tension, and imagining a pleasant and relaxing image.

Let us begin with the breathing. Because of the pace of our daily life, we learn to breathe more quickly than we need to. We can slow down our respiration cycle without any serious side effects. The cycle I use in teaching people runs on five-second intervals. Five seconds to inhale a full breath using the *whole lung*. Five seconds of holding that air. Five seconds of exhaling the air in a slow steady column. Five seconds of sitting quietly without drawing your next breath. Then the twenty-second cycle begins again. Try it right now. Like any other skill, learning this will take practice. Go at your own pace, but leave time for each of the four intervals, and go more slowly than you normally would. (If you are a smoker, you will have more difficulty with these exercises.) Five seconds: inhale. Five seconds: hold. Five seconds: exhale. Five seconds: hold. Then begin again.

The second part of these exercises includes the release of muscle tension in the muscle groups listed below. Our minds are remarkably powerful instruments for coping with stress. If you think of your muscles being freed from tension, your brain will, in fact, release the muscle tension. When you are breathing slowly as outlined above, think of the various muscles in your body to be freed of tension. Think of their being released, and they will be. For example, inhale, hold (release the muscle tension in your toes, arches, and heels), exhale, hold. Inhale, hold (release the muscle tension in your ankles, shins, and knees), exhale, hold, and so forth.

Below is a list of the muscle clusters that I teach to others. Do one grouping at a time, and remember to maintain your slower breathing pace.

1. Toes, arches, and heels;
2. Ankles, shins, and knees;
3. Thighs, buttocks, and anal sphincter;
4. Lower back, up the back to the neck and shoulders;
5. Abdomen, chest muscles, again up to the neck and shoulders;
6. Upper arms, forearms, down to the wrists;
7. Each hand, each finger, each fingertip;
8. The muscles in each shoulder and all around the base of the neck;
9. The whole neck, the tongue, and the jaw;
10. The dental cavity, the mouth muscles, and the upper facial cheeks;
11. The eye muscles, the forehead, and the top and back of the scalp.

Breathe slowly, and go through each muscle grouping—one grouping per respiration cycle after you have inhaled. You can complete these two parts of the relaxation exercise in about ten to fifteen minutes after you have had a chance to practice.

The third component of the approach is the addition of a pleasant and relaxing image. Think of some place you have been that was pleasant or some place that you would like to visit. Make sure the place you choose has no unpleasant memories for you. Do not select that secluded beach where you broke up with your sweetheart, do not select that breathtaking mountain where you broke your leg skiing. Make sure your choice is truly pleasant for you.

After you are breathing slowly and rhythmically, and have released all the muscle tension in your body, close your eyes and imagine your special place. Make your image as real as if you were actually there. Whatever you might see or taste or touch or smell if you were really there, be sure to include those things in your image. If your image is unclear, or if you have trouble thinking in images, picture yourself floating on a cloud of your favorite color, or picture yourself sitting before a curtain blowing in the summer breezes. Imagine your pleasant scene for five minutes. Be sure you continue to breathe slowly, and keep all the muscle tension in your body released. Relax.

With a little practice, these relaxation exercises of breathing slowly, releasing muscle tension, and imagining a relaxing place can be quite effective in reducing your stress response. If you have a particularly stressful day, you can do them for ten minutes in the morning, ten minutes at noon, ten minutes at suppertime, and ten minutes before bedtime. These exercises have the added advantage of being flexible. If you are anticipating some event which is making you anxious, these exercises can help beforehand. You can also use them when the event is over to return your body to its normal stress-free resting state. These exercise components take practice, but they are very helpful in reducing the negative fallout from general life stress and from the specific heightened distress often found in victims of violence and abuse.

APPENDIX C

Thoughts for Caregivers

There is increasing opportunity for counselors to appreciably reduce the suffering associated with posttraumatic stress disorder. Many persons are now realizing that they are victims, and they are reaching out for help. Simultaneously, medical and behavioral science is rapidly improving our ability to treat such victims, and this book has presented many of the most recent findings.

All of us as counselors have two important tasks as we go about our work ministering to trauma victims. The first is to be as professionally well-versed as we can be, and the second is to limit in ourselves any possible effects of vicarious traumatization as we go about our work. Both are equally important.

Professional Training. The first basic requirement for any of us is that we be competent, trained, general practice counselors. As with other aspects of mental health services, there is no substitute for a good therapist. In addition to this, most counselors find it helpful to have some specific training in PTSD. Unless you are a recent graduate, you probably did not have such formal training in treating victims of psychological trauma. It is important to acquire the training as it greatly enhances the quality of care that you will ultimately provide. Graduate schools now offer such courses, and continuing education programs are also now available. Any of us, of course, can arrange for individual or peer supervision.

Psychiatrist Frank Putnam (1989) has addressed some of the common concerns that counselors have when they work with victims of trauma. Some counselors feel that these patients need to be treated primarily in hospital settings. This is rare. While an occasional brief admission may prove helpful in managing suicidal thoughts or in containing excessive physiological arousal and/or intrusive memories, long-term inpatient care is usually not the treatment of choice. Most counselors have access to a physician who can prescribe medicines, and arrange for such brief admissions, if need be. With these safeguards in place, treating victims of traumatic events is similar to counseling persons with other types of diagnostic problems.

Another common therapist concern is fear of eliciting an angry, rageful state in the victim that will lead to management problems. If the victim's treatment proceeds in recovering memories in small, manageable steps, this is unlikely to happen. Even when such anger is expressed, most victims remain in control, and can curtail the expression of this anger within short periods of time.

Members of the counseling community may also have concerns about the role of medicines as an adjunctive treatment; the best approach to the family in cases of incest and domestic violence; how to reach difficult families where one or both parents are cruel, addicted, have abused many family members, or refuse treatment (Forward, 1989). These are reasonable concerns, but much technical assistance is available, and counselors need not be unduly concerned about inclusion of PTSD victims in their practices.

Managing Vicarious Traumatization. We counselors need to be mindful that dealing with victims of trauma and PTSD may place us at risk for increased personal distress (McCann and Pearlman, 1990). In one-to-one counseling, we may find that repeated victim themes of abuse may increase in us therapists concerns with safety, trust, and vulnerability. Likewise, emergency services personnel may be at increased risk for developing PTSD because many of them are actual witnesses to traumatic events. A recent study of crisis workers found that over eighty-five percent of the police, fire, nurse, and EMT personnel who were interviewed had symptoms of psychological trauma. The death of a child, a rescue event that may threaten the life of the rescuer, the anguish and suffering of others, and dealing with the media at the scene may produce distress in any of us. Counselors too can end up with sleep disturbances, flashbacks, and a preoccupation with death. Developing addictive behaviors is no more helpful to us than it is to victims. There are strategies that counselors and emergency services personnel can employ in their own work to minimize such personal disruptions and to enhance the quality of care they provide, and they should be utilized. Here are some that have proven helpful.

Individual Counselors. It is important that counselors manage the daily life stress in their own personal lives adaptively (Flannery, 2003). This ensures that the continuous, professional involvement in the lives of victims will not be an undue burden. Individual or peer supervision is another resource that we have mentioned. In addition, it is important for all counselors to have a professional support network where such therapist-victim issues can be addressed. This support will provide us with perspective and enhance the quality of care that we deliver.

Emergency Services Personnel. Emergency service workers are the frontline response to the community's worst carnage. The events that you must respond to are highly stressful. In addition to generally learning how to manage stress in the ways noted above, emergency service workers will find further helpful suggestions in the available research.

There is no substitute for pre-incident training as a method for reducing unnecessary stress. The more extensive the training has been, the easier it will be for the rescuer to help at the site. At the site, the emergency services personnel should train themselves to continuously gather information, remain part of the team effort, do relaxation exercises as time permits (appendix B), and seek support. Rescue work is hard physical labor and mentally very taxing. Two hours on and thirty minutes off appears to be the most optimal way to cope at the site. When the event is over, rescue workers should exercise, eat, rest regular hours, and avoid boredom.

In addition, we are learning that Critical Incident Stress Debriefing, which has proven to be a helpful intervention for victims, is equally helpful for caretakers.

Part of my recent professional responsibilities has been to design and implement an occupational trauma program for mental hospital nursing personnel who are at risk for assault by patients in the course of their work site responsibilities. We call our program ASAP, the Assaulted Staff Action Program. My colleagues and I provide a system-wide, twenty-four-hour on-call crisis intervention for over four hundred staff at several patient-care sites (Flannery, 1998).

When a staff member has been assaulted, an ASAP team member responds within ten minutes to provide crisis intervention debriefing. The team member focuses on what has happened, how the victim is feeling, and works with the victim to begin to restore mastery, attachment, and meaning. While the program responds primarily to assaultive episodes, we have responded to cases of sudden patient death, the homicide of a staff member on the way to work, serious suicide attempts by patients, and the like.

This program responds to individual provider needs. It appears helpful in treating the unwanted consequences of psychological trauma and is one method of containing industrial accident costs. This approach can be easily adapted to the needs of police, fire, emergency medical technicians, correction facilities, and school systems where individuals may be victims of or witnesses to traumatic events.

Emergency services personnel will greatly benefit from reduced stress if crisis debriefing is done immediately after the specific disaster the personnel have responded to. An occupational trauma program such as ASAP is most helpful when it is continuously available for the needs of staff.

References

Flannery, R. B., Jr. *The Assaulted Staff Action Program (ASAP): Coping with the Psychological Aftermath of Violence.* Ellicott City, MD: Chevron Publishing, 1998.

McCann, L., and Pearlman, L.A. *Psychological Trauma and the Adult Survivor: Theory, Therapy, and Transformation.* New York: Brunner/Mazel, 1990.

Select Readings

Chapter 1: What Has Befallen Me?
Psychological Trauma and Posttraumatic Stress Disorder

Braun, B.G., ed. *Treatment of Multiple Personality Disorder.* Washington, DC: American Psychiatric Press, 1986.

Danieli, Y. "Mourning in Survivors and Children of Survivors of the Nazi Holocaust: The Role of Group and Community Modalities." In Dietrich, D., and Shabad, P., eds. *The Problem of Loss Mourning: Psychoanalytic Perspectives.* New York: International Universities Press, 1989. 427-460.

Figley, C.R., ed. *Trauma and Its Wake. Vol. II: Traumatic Stress, Theory Research, and Intervention.* New York: Brunner/Mazel, 1985.

Flannery, R.B., Jr. *Becoming Stress-Resistant Through the Project SMART Program.* Ellicott City, MD: Chevron Publishing, 2003.

Flannery, R.B., Jr., and Harvey, M.R. "Psychological Trauma and Learned Helplessness: Seligman's Paradigm Reconsidered." *Psychotherapy* 28 (1991):374-378.

Horowitz, M.J. *Stress Response Syndromes.* 2nd Edition. Northvale, NJ: Aronson, 1986.

Kukla, R.A., Schlenger, W.E., Fairbank, J.A., Hough, R.L., Jordan, B.K., Marmar, C.R., Weiss, D.S., and Chapter by Grady, D.A. *Trauma and the Vietnam War Generation: Report of Findings from the National Vietnam Veterans Readjustment Study.* New York: Brunner/Mazel, 1990.

Lavelle, J., and Mollica, R. "Southeast Asian Refugees." In Comas-Dias, L., and Griffith, E.E., eds. *Clinical Guidelines in Cross-Cultural Mental Health.* New York: Wiley and Sons, 1985. 262-304.

Litz, B.T., and Keane, T.M. "Information Processing in Anxiety Disorders: Application to the Understanding of Post-Traumatic Stress Disorder." *Clinical Psychology Review,* 9 (1989):243-257.

Putnam, F.W., Jr. *Diagnosis and Treatment of Multiple Personality Disorder.* New York: Guilford Press, 1989.

Seligman, M.E.P. *Helplessness: On Depression, Development, and Death.* New York: W.H. Freeman, 1975.

van der Kolk, B.A., ed. *Psychological Trauma.* Washington, DC: American Psychiatric Press, 1987.

Chapter 2: Am I Losing My Mind?
The Psychology of Posttraumatic Stress Disorder

Antonovsky, A. *Health, Stress, and Coping.* San Francisco: Jossey-Bass, 1979.

Beck, A.T. *Depression: Causes and Treatment.* Philadelphia: University of Pennsylvania Press, 1970.

Becker, E. *The Denial of Death.* New York: Free Press, 1973.

Ellis, A. *Rational-Emotive Psychotherapy.* New York: Stuart, 1963.

Goleman, D. *Vital Lies, Simple Truths: The Psychology of Self-Deception.* New York: Simon and Schuster, 1985.

Hawton, K., Salkovskis, P.M., Kirk, J., and Clark, D. *Cognitive Behaviour Therapy for Psychiatric Problems: A Practical Guide.* New York: Oxford University Press, 1989.

Janoff-Bulman, R. "The Aftermath of Victimization: Rebuilding Shattered Assumptions." In Figley, C.R., ed. *Trauma and Its Wake: The Study and Treatment of Post-traumatic Stress Disorder.* New York: Brunner/Mazel, 1985, 15-35.

Kübler-Ross, E. *On Death and Dying.* New York: Macmillan, 1969.

Lindemann, E. "Symptomatology and Management of Acute Grief." *American Journal of Psychiatry,* 101 (1944):141-148.

Taylor, S.E. *Positive Illusions: Creative Self-Deception and the Healthy Mind.* New York: Basic Books, 1986.

Tenner, H., and Affleck, G. "Blaming Others for Threatening Events," *Psychological Bulletin,* 108 (1990):209-232.

Wortman, C.B., and Silver, R.C. "The Myths of Coping with Loss," *Journal of Consulting and Clinical Psychology,* 57 (1989):349-357.

Chapter 3: Why Are My Nerves So Frayed?
The Biology of Posttraumatic Stress Disorder

Cannon, W. *The Wisdom of the Body.* New York: Norton, 1963.

Fromm, E. *The Heart of Man: Its Genius for Good and Evil.* New York: Harper and Row, 1964.

Fromm, E. *The Anatomy of Human Destructivenss.* New York: Holt, Rinehart, and Winston, 1973.

Giller, E.L., Jr., ed. *Biological Assessment and Treatment of Post-traumatic Stress Disorder.* Washington, DC: American Psychiatric Press, 1990.

Herman, J.L. *Trauma and Recovery: The Aftermath of Violence—From Domestic Abuse to Political Terror.* New York: Basic Books, 1992.

Lazarus, R.S., and Folkman, S. *Stress, Appraisal, and Coping.* New York: Springer, 1984.

Litz, B., Gray, M., and Adler, A. "Early Intervention for Trauma: Current Status and Future Directions." *Clinical Psychology: Science and Practice,* 9 (2002) 112-137.

McGaugh, J.L. "Significance and Remembrance: The Role of Neuromodulatory Systems." *Psychological Science,* 1 (1990):15-25.

Selye, H. *The Stress of Life.* New York: McGraw-Hill, 1956.

van der Kolk, B.A., McFarlane, A.C., and Weisaeth, L. (eds.): *Traumatic Stress: The Effects of Overwhelming Experience on Mind, Body and Society.* New York: Guilford, 1996.

Widom, C.S. "Does Violence Beget Violence? A Critical Examination of the Literature." *Psychological Bulletin,* 106 (1989):3-28.

Wilson, J.P. *Trauma, Transformation, and Healing: An Integrative Approach to Theory, Research, and Post-Traumatic Therapy.* New York: Brunner/Mazel, 1989.

Chapter 4: It Hurts. The Several Faces of Untreated Trauma

Bushman, B. J., and Cooper, H.M. "Effects of Alcohol on Human Aggression: An Integrative Research Review" *Psychological Bulletin,* 107 (1990):341-354.

Carmen, E.H., Reiker, P.P., and Mills, T. "Victims of Violence and Psychiatric Illness." *American Journal of Psychiatry,* 141 (1984):378-383.

Fish-Murray, C.C., Koby, E.V., and van der Kolk, B.A. "Evolving Ideas: The Effect of Abuse on Children's Thought." In van der Kolk, B.A., ed. *Psychological Trauma.* Washington, DC: American Psychiatric Press, 1987, 89-110.

Ford, C.V. "The Somatizing Disorders." *Psychosomatics,* 27 (1986):327-337.

Herman, J.L., Perry, J.C., and van der Kolk, B.A. "Childhood Trauma in Borderline Personality Disorder." *American Journal of Psychiatry,* 146 (1989): 490-495.

Jones, J.C., and Barlow, D.H. "The Etiology of Post-Traumatic Stress Disorder." *Clinical Psychology Review,* 10 (1990):299-328.

Justice, B. *Who Gets Sick: How Beliefs, Moods, and Thoughts Affect Your Health.* Los Angeles: J.P. Tarcher, 1988.

Khantzian, E. "The Self-Medication Hypothesis of Affective Disorders: Focus on Heroin and Cocaine Addiction" *American Journal of Psychiatry*, 142 (1985):1259-1264.

Kluft, R.P., ed. *Incest-Related Syndromes of Adult Psychopathology.* Washington, DC: American Psychiatric Press, 1990.

Lynch, J.R. *The Broken Heart: The Medical Consequences of Human Loneliness.* New York: Basic Books, 1977.

McNally, R. *Remembering Trauma,* Cambridge, MA: Harvard University Press, 2003.

Watson, D., and Clark, L.A. "Negative Affectivity: The Disposition to Experience Aversive Emotional States." *Psychological Bulletin* 96 (1985):465-490.

Chapter 5: Sexual Abuse

Barry, K. *Female Sexual Slavery.* New York: New York University Press, 1979.

Brownmiller, S. *Against Our Will: Men, Women and Rape.* New York: Simon and Schuster, 1975.

Burgess, A.W., and Holstrom, L.L. "Rape Trauma Syndrome." *American Journal of Psychiatry* 131 (1974):981-985.

Everly, G.S., Jr., and Lating, J.R. *Psychotraumatology: Key Papers and Care Concepts in Post-Traumatic Stress.* New York: Plenum, 1995.

Gelinas, D. J. "The Persisting Negative Effects of Incest." *Psychiatry* 46 (1983):312-332.

Hartman, C.R., and Burgess, A.W. "Rape Trauma and Treatment of the Victim." In Ochberg, F.M., ed. *Post-Traumatic Therapy and Victims of Violence.* New York: Brunner/Mazel, 1988, 152-174.

Herman, J.L., with Hirshman, L. *Father-Daughter Incest.* Cambridge, MA: Harvard University Press, 1981.

Lew, M. *Victims No Longer: Men Recovering from Incest and Other Sexual Child Abuse.* New York: Harper and Row, 1988.

Russell, D.E.H. *The Secret Trauma: Incest in the Lives of Girls and Women.* New York: Basic Books, 1986.

Russell, D.E.H. *Rape in Marriage.* 2nd ed. Bloomington, IN: Indiana University Press, 1990.

Sanday, P.R. *Fraternity Gang Rape: Sex, Brotherhood, and Privilege on Campus.* New York: New York University Press, 1990.

Warshaw, R. *I Never Called It Rape: The Ms. Report on Recognizing, Fighting, and Surviving Date and Acquaintance Rape.* New York: Harper and Row, 1988.

Chapter 6: Physical Abuse

Blackman, J. *Intimate Violence: A Study of Injustice.* New York: Columbia University Press, 1989.

Browne, A. *When Battered Women Kill.* New York: Free Press, 1987.

Gelles, R.J., and Strauss, M.A. *Intimate Violence: The Definitive Study of the Causes and Consequences of Abuse in the American Family.* New York: Simon and Schuster, 1988.

Gillespie, C.K. *Justifiable Homocide: Battered Women, Self-Defense and the n Law.* Columbus, OH: Ohio State University Press, 1989.

Justice, B., and Justice, R. *The Abusing Family.* New York: Human Sciences Press, 1976.

Kempe, R.S., and Kempe, C.H. *Child Abuse.* Cambridge, MA: Harvard University Press, 1978.

Lifton, R.J. *Boundaries: Psychological Man in Evolution.* New York: Random House, 1969.

Sarrell, P.M., and Masters, W.H. "Sexual molestation of men by women." *Archives of Sexual Behavior* 11 (1982):117-121.

Stacey, W., and Shupe, A. *The Family Secret: Domestic Violence in America.* Boston: Beacon Press, 1983.

Straus, M.A., Gelles, R.J., and Steinmetz, S.K. *Behind Closed Doors: Violence in the American Family.* New York: Anchor Books, 1980.

Walker, L. *The Battered Woman.* New York: Harper and Row, 1979.

Walker, L. *The Battered Woman Syndrome.* New York: Springer, 1984.

Chapter 7: Combat / Family Alcoholism

Beattie, M. *Co-Dependent No More: How to Stop Controlling Others and Start Caring for Yourself.* New York: Harper/Hazelden, 1987.

Bradshaw, J. *Healing the Shame that Binds You.* Deerfield Beach, FL: Health Communications, 1988.

Dimsdale, J.E., ed. *Survivors, Victims, and Perpetrators: Essays on the Nazi Holocaust.* New York: Hemisphere Publishing, 1980.

Emery, V.O., Emery, P.E., Shama, D.K., Quiana, N.A., and Jassami, A.K. "Predisposing Variables in PTSD Patients." *Journal of Traumatic Stress* 4 (1991):325-343.

Erikson, E.H. *Childhood and Society.* 2nd ed. New York: W.W. Norton, 1963.

Figley, C.R., ed. *Stress Disorders Among Vietnam Veterans: Theory, Research, and Treatment Implications.* New York: Brunner/Mazel, 1978.

Forward, S., with Buck, C. *Toxic Parents: Overcoming Their Hurtful Legacy and Reclaiming Your Life.* New York: Bantam, 1989.

Fossum, M.A., and Mason, M.J. *Facing Shame: Families in Recovery.* New York: W.W. Norton, 1986.

Lifton, R.J. *Death in Life: Survivors of Hiroshima.* New York: Random House, 1967.

Lynch, J. *A Cry Unheard: New Insights into the Medical Consequences of Loneliness.* Baltimore: Bancroft, 2000.

Wilson, J.P., Hazel, Z., and Khana, B., eds. *Human Adaptation to Extreme Stress: From the Holocaust to Vietnam.* New York: Plenum Press, 1988.

Woititz, J.G. *Adult Children of Alcoholics.* Hollywood, FL: Health Communications, 1983.

Chapter 8: Recovery: Help for Victims

Flannery, R.B., Jr. "From Victim to Survivor: A Stress Management Approach in the Treatment of Learned Helplessness." In van der Kolk, B.A., ed. *Psychological Trauma.* Washington, DC: American Psychiatric Press, 1987, 217-232.

Flannery, R.B., Jr. "Psychological Trauma and Posttraumatic Stress Disorder: A Review." *International Journal of Emergency Mental Health* 1 (1999):135-140.

Fox, S.S., and Scherl, D.J. "Crisis Intervention With Victims of Rape." *Social Work* 17 (1972):37-42.

Herman, J.L., and Schatzow, E. "Time-Limited Group Therapy for Women with a History of Incest." *International Journal of Group Psychotherapy* 34 (1984): 605-616.

Kasl, C.D. *Women, Sex, Addiction: A Search for Love and Power.* New York: Ticknor and Fields, 1989.

Kempe, C.H., Silverman, F.N., and Steele, B.J. "The Battered Child Syndrome," *Journal of the American Medical Association* 181 (1962):17-24.

Koss, M., and Harvey, M.R. *The Rape Victim: Clinical and Community Approaches.* rev. ed. San Francisco: Sage, 1991.

Kushner, H.S. *When Bad Things Happen to Good People.* New York: Schocken Books,1981.

Lerner, R. *Daily Affirmations: For Adult Children of Alcoholics.* Deerfield, FL: Health Communications, 1985.

Ogden, G. *Sexual Recovery: Everywoman's Guide Through Sexual Co-Dependency.* Deerfield, FL: Health Communications, 1990.

Pennebaker, J.W. *Opening Up: The Healing Power of Confiding in Others.* New York: Morrow, 1990.

Siegel, D.J.: *The Developing Mind: Toward A Neurobiology of Interpersonal Experience.* New York: Guilford, 1999.

Chapter 9: Recovery: Help from Families and Friends

Bass, E., and Davis, L. *The Courage to Heal: A Guide for Women Survivors of Child Abuse.* New York: Harper and Row, 1988.

Bouza, A.V. *The Police Mystique: An Insider's Look at Cops, Crime, and the Criminal Justice System.* New York: Plenum, 1990.

Edwards, S.M. *Policing Domestic Violence: Women, the Law, and the State.* San Francisco: Sage, 1990.

Figley, C.R. "Post-traumatic Family Therapy." In Ochberg, F.M. (Ed.), *Post-Traumatic Therapy and Victims of Violence.* New York: Brunner/Mazel, 1989, 83-109.

Fletcher, C. *What Cops Know: Cops Talk About What They Do, How They Do It, and What It Does to Them.* New York: Villard Books, 1991.

Havens, L. *Making Contact: Use of Language in Psychotherapy.* Cambridge, MA: Harvard University Press, 1986.

Kritzberg, W. *The Adult Children of Alcoholics: From Discovery to Recovery.* New York: Bantam, 1988.

Krugman, S. "Trauma in The Family: Perspectives on the Intergenerational Transmission of Violence." In van der Kolk, B.A., ed. *Psychological Trauma.* Washington, DC: American Psychiatric Press, 1987, 127-151.

Mitchell, J.T., and Everly, G.S., Jr. *Critical Incident Stress Debriefing: An Operations Manual for the Prevention of Traumatic Stress Among Emergency Services and Disaster Workers.* Ellicott City, MD: Chevron Publishing, 1996.

Raphael, B. *When Disaster Strikes: How Individuals and Communities Cope with Disaster.* New York: Basic Books, 1986.

Rocklin, R., and Lavett, D.K. "Those Who Broke the Cycle: Therapy with Nonabusive Adults Who Were Abused as Children." *Psychotherapy: Theory, Research and Practice* 24 (1987):769-778.

Tavris, C. *Anger: The Misunderstood Emotion.* New York: Simon and Schuster, 1982.

Chapter 10: Why Me?
Searching for the Soul of Goodness in Things Evil

Allport, G.W. *The Nature of Prejudice*. Garden City, NY: Doubleday/Anchor, 1958.

American Psychiatric Association. *Diagnostic and Statistical Manual of Mental Disorders*. Fourth Edition. Washington, DC: American psychiatric Press, 1994.

de Chardin, T. *The Phenomenon of Man*. New York: Harper and Row, 1959.

Durant, W., and Durant, A. *The Lessons of History*. New York: Simon and Schuster, 1968.

Hick, J. *Evil and the God of Love*. New York: Harper and Row, 1966.

James, W. *Varieties of Religious Experience*. New York: New American Library, 1958.

Lewis, C.S. *The Problem of Pain*. New York: Macmillan, 1962.

Mayer, K.E. *The Psychobiology of Aggression*. New York: Harper and Row, 1976.

Menninger, K. *Whatever Became of Sin?* New York: Hawthorn Books, 1973.

Peck, M. S. *People of the Lie: The Hope for Healing*. New York: Simon and Schuster, 1983.

Sanford, N., and Comstock, C., eds. *Sanctions for Evil: Sources of Social Destructiveness*. Boston: Beacon Press, 1971.

van Kaam, A. *Envy of Originality*. Garden City, NY: Doubleday, 1972.

Acknowledgments

Every reasonable effort has been made to locate the owners of rights to previously published material printed herein. The author and publisher gratefully acknowledge permission from the following sources to print material in this book:

Chart from *The Battered Woman* by Lenore E. Walker. Copyright © 1979 by Lenore E. Walker. Reprinted by permission of HarperCollins Publishers and Mary Yost Associates, Inc.

Chart from *I Never Called It Rape: The MS. Report on Recognizing, Fighting and Surviving Date and Acquaintance Rape* by Robin Warshaw. Copyright © 1988 by The MS. Foundation for Education and Communication, Inc. and Sarah Lazin Books. Reprinted by permission of HarperCollins Publishers.

1 Corinthians 13 verses, 1-8 and 13 from the Revised Standard Version of the Bible, copyright ©1946, 1952, 1971 by the Division of Christian Education of the National Council of the Churches of Christ in the USA and used by permission.

Index

225

About the Author

Raymond B. Flannery, Jr., Ph.D., a licensed clinical psychologist, is Associate Clinical Professor of Psychology, Department of Psychiatry, Harvard Medical School and Adjunct Assistant Professor of Psychiatry, Department of Psychiatry, The University of Massachusetts Medical School. An internationally recognized expert, Dr. Flannery has lectured extensively in Canada, Europe, and the United States and is the author of six books and over 100 peer-reviewed articles in medical and scientific journals on the topics of stress, violence, and victimization. His work has been translated into four foreign languages.

Dr. Flannery and his wife live in the suburbs of Boston.